The Big Dig

NEW ENGLAND REMEMBERS

The Big Dig

James A. Aloisi, Jr.

CE

Commonwealth Editions
Beverly, Massachusetts

Library of Congress Cataloging-in-Publication Data
Aloisi, James A.
The Big Dig / James A. Aloisi, Jr.
p. cm. — (New England remembers)
Includes index.
ISBN 1-889833-82-7
1. Central Artery/Third Harbor Tunnel Project (Mass.)—History. 2. Tunnels—Massachusetts—Boston—History. 3. Express highways—Massachusetts—Boston—History. I. Title. II. Series.
HE356.5.B6A76 2004
388.1'3—dc22 2004012814

Cover and interior design by Laura McFadden.
laura.mcfadden@rcn.com

Printed in the United States

Published by Commonwealth Editions
an imprint of Memoirs Unlimited, Inc.
266 Cabot Street, Beverly, Massachusetts 01915
www.commonwealtheditions.com.

New England Remembers series
Robert J. Allison, Series Editor
The Hurricane of 1938
The Big Dig

All photos are courtesy of the Massachusetts Turnpike Authority, Central Artery/Tunnel Project.

The "New England Remembers" logo features a photo of the Thomas Pickton House, Beverly, Massachusetts, used courtesy of the Beverly Historical Society.

CONTENTS

To my father, who bought me the Golden Book Encyclopedia and began my lifelong love affair with history

FOREWORD

IT IS ONE OF THE LARGEST public works projects in the history of the country. Boston's Central Artery/Tunnel Project built a new underground highway system beneath one of the oldest cities in North America, underneath a rusty and deteriorating elevated highway built fifty years earlier. This was accomplished without closing the old highway, which carried 190,000 cars every day, and without taking a single home. Though Bostonians sometimes grumbled about the inconvenience during construction, and though costs escalated beyond the wildest estimates, when the new underground highway opened and the old eyesore came down, all marveled at the way the city had been transformed.

This is the story of the Big Dig, told by James A. Aloisi, Jr. Jim Aloisi has a unique perspective—he was Assistant Secretary of Transportation in Governor Michael Dukakis's administration, when the plans were made to undertake this massive project, and he was general counsel to the Massachusetts Turnpike Authority during much of the early construction. He has an insider's insight, a historian's judgment, and a storyteller's narrative grace.

The Big Dig is the second in a new series of books—New England Remembers—about people and events that have shaped New England's history. Written with economy and insight, these books capture the unique character of New England. The Big Dig required as much political skill as engineering expertise to pull off, and in the end transformed Boston in ways few could have imagined. New England will long remember the Big Dig.

Robert J. Allison, Series Editor
Boston, Massachusetts

INTRODUCTION

THIS IS A SHORT HISTORY of the completion of Interstate Highways 93 and 95 in Boston, Massachusetts—the Central Artery/Tunnel Project, commonly known as the "Big Dig." A more detailed study of the Project could—and should—be written. Readers searching for every detail and looking to identify every person integral to the Project will have to wait for a much larger, comprehensive account.

As Assistant Transportation Secretary in the late 1980s, and later as General Counsel to the Massachusetts Turnpike Authority from March 1989 to June 1996, I was a witness to or participant in a number of the events described herein. This experience enables me to write about certain aspects of the Big Dig from a "behind-the-scenes" perspective that I hope the reader will find interesting and informative. Nevertheless, this telling of the Big Dig story must fairly be read as my own unique and personal point of view.

PRELUDE

The Past Is Prologue

THE MANSION HOMES OF BOSTON'S BACK BAY have always been status symbols. As Boston grew from a small town into a modern city, the Back Bay became its most desirable neighborhood—an elegant expression of Boston's desire, according to the architectural historian Bainbridge Bunting, "to assume a place among the great cities of the world." Urban planner and critic Lewis Mumford considered the Back Bay, with its tree-lined avenues and grand and spacious street grid, one of the outstanding American city-planning achievements of the nineteenth century.

The cachet of the Back Bay began almost immediately after its reclamation from the large tidelands that marked the original topography of Boston. In a famous scene in John Marquand's *The Late George Apley*, George Apley recalls that his father, Thomas, wasted no time deciding that he must move his family out of the South End and to the Back Bay because he observed a man in shirtsleeves standing on the stairs of a nearby brownstone. Such a sight meant that the South End was not a proper place for a proper Bostonian. The South End, built as a fashionable alternative to the cramped and tired Beacon Hill, gave way to the newer and more fashionable Back Bay, where no one would be seen in shirtsleeves.

In the late nineteenth century and through a good part of the next, much of Boston's ruling class inhabited the Back Bay. It was home to families named

Storrow, Ames, and Coolidge, to Governors John Andrew and Alvan Fuller, and, before she moved to her famous Italianate villa in the Fenway, to Isabella Stewart ("Mrs. Jack") Gardner. It was a grand and gracious place, adjacent to the Charles River, modern and decidedly European in flavor—from the great mall extending down Commonwealth Avenue to the very names of the cross streets, evoking English places—Berkeley, Clarendon, Dartmouth, Exeter, Gloucester, Hereford.

A home on the water side of Beacon Street was prized by those who could afford to build there. "There ain't a sightlier place in the world for a house" exclaimed William Dean Howells's protagonist Silas Lapham, as he urged his wife to consider moving from their South End residence to one in the new Back Bay. The Back Bay was *the* place to be—except during low tide. The ebbing tide on the Charles revealed a gray mud flat, which emitted foul odors throughout the Back Bay. At low tide, the Charles was not an attractive public amenity. In a word, it stank.

Today's Charles River Esplanade—a magnet for walkers, runners, sunbathers, and others seeking a tranquil and attractive respite from the busy city—and the river itself, filled with sailboaters, scullers, and tourist boats, are in sharp contrast to the river in the late nineteenth century. Then the river was largely inaccessible to most citizens, and its sewage-filled mud flats did not attract even the hardiest of souls. The Charles River was not serving the needs of the city or its people, and it posed a significant threat to Boston's aspirations to greatness. Something had to be done about it.

Charles Eliot, the son of the president of Harvard University, was determined to make a contribution to the improvement of his community. Eliot apprenticed in Frederick Law Olmsted's Brookline office before opening his own landscape architecture firm. He put forward an ambitious agenda for parks improvements, not just in Boston, but throughout the metropolitan area. Eliot had a vision of a Charles River Basin, bordered by a grand promenade, as a "central court of honor" in the city. Eliot died at an early age but his cause was taken up by other young civic leaders—most notably James Jackson Storrow.

Harvard-educated, athletic, with an optimistic and forward-looking vision for the city, Storrow grew up in a riverside townhouse on Beacon Street and was captain of the Harvard varsity crew in 1885. He knew how smelly and ugly the river was at low tide, and he saw the possibility of transforming the river into a prime recreational asset.

Storrow led the effort to build a dam at Craigie Bridge and create a tideless basin, eliminating the stench and transforming the Charles into an important recreational amenity for all citizens. He was joined by Charles Eliot's father, the Harvard president, who described the push to improve the Charles River as a natural and proper mission for the public sector: "municipalities [do not] exist for profit in money, but for the people who live in them, and the supreme object of any city should be the well-being and happiness of the community."

Many Back Bay landowners opposed the drive to create the Charles River Basin. Though damming the river would eliminate the foul stench of low tide, beautify the river embankment, and make efficient navigation possible during all hours of the day, these Beacon Street neighbors opposed the plan because they were, to their core, Bostonians—and Bostonians do not accept change easily. Marquand's fictional Thomas Apley complained that creating the Charles River Basin was "an infringement on owners' rights [undertaken] to gratify the unbalanced whim of a small group who believe there are not sufficient places for the citizens of Boston to walk and play. It is not the purpose of those who built upon the Charles River to have playgrounds in their back yards. The Boston Common was intended for recreation, and also the Public Gardens; and these generous contributions to the city's welfare are enough."

In the end, Apley's cramped view of progress did not carry the day. Storrow prevailed, and the Charles River Basin has been an invaluable recreational amenity for over a century. Today we take for granted what required the hard work and dedication of men of vision, commitment, and perseverance—forward-looking people who understood that Boston is a special place on a constrained peninsula, and that it is important to find ways to maximize that precious acreage for the benefit of its citizens.

Such men—and women—are hard to find and do not often rise to the occasion. But in the late twentieth century, Boston was blessed with civic leaders who, for the most part, committed themselves to another massive public improvement. This time, the city was presented with the opportunity to rebuild its downtown core—an opportunity rarely given to cities 370 years old.

Downtown Boston was a victim of poor transportation planning. The Central Artery, an elevated highway built in an era when planners thought nothing of tearing apart the neighborhoods to advance "progress," quickly became a congested, inefficient answer to the problem of mobility through the urban core. Designed at midcentury to reinvigorate the city, the Central Artery twenty years later was causing blight and stagnation and, although it did not stink at low tide, the Central Artery was as ugly and offensive as the Charles River tidelands ever were.

This is the story of how Boston rebuilt itself in the last decade of the twentieth century through an enormous highway-expansion project. Affectionately called the "Big Dig," it was one of the nation's greatest urban transportation infrastructure projects. It also included a large public investment in urban environmental remediation and reclamation, establishing new opportunities for the city that, as of this writing, are impossible to fully predict.

This is also the story of the civic and political leaders who had a vision, made that vision a possibility, and turned it into reality. It might have been a lot less expensive, as Congressman Barney Frank famously observed, to raise the city rather than lower the Central Artery, but that would never have been an elegant or proper solution—certainly not a solution worthy of the Hub of the Universe.

CHAPTER ONE

The Central Artery: Boston's Failed Experiment

AT THE MIDPOINT of the twentieth century, Boston was at a crossroads. For too many years city leaders stood by as Boston gradually lost its place as a great urban center. Its industrial base had deteriorated and the once vibrant seaport began to stagnate as business left for New York's better, busier port. The city's neighborhoods, depressed by lack of investment and torn apart by insensitive transportation projects like the Sumner Tunnel and the Washington Street elevated transit line, became places people longed to escape from.

In 1950 the election of a new mayor presented city voters with a choice between the legendary James Michael Curley and the former city clerk, quiet and bespectacled John Hynes. Hynes had served as acting mayor when Curley was doing time in federal prison. When Curley returned to reclaim his job, he rudely pushed the hard-working Hynes aside, brusquely remarking that he had accomplished more in an afternoon than Hynes had in six months. Hynes would not soon forget the insult, and the 1950 election offered him the chance to exact revenge on the aging and increasingly out of touch Curley. Hynes won decisively, and his victory marked the beginning of a new era in the city's history—the beginning of the "New Boston."

Hynes was committed to urban renewal and change. Time was quickly passing Boston by, and unless the city moved quickly and decisively, it would forever be an urban backwater. Working closely with the business community and with civic leaders like W. Seavey Joyce, dean of Boston College Business School (and later Boston College president), Hynes set in motion a plan and a vision for the New Boston that was meant to be a model of twentieth-century progress. The New Boston was shorthand for a variety of exciting initiatives—the 135-mile Massachusetts Turnpike, the Prudential Center, the razing of the West End to create a modern "Government Center," and the construction of a second harbor tunnel. Hynes pushed hard and successfully for legislation providing important tax breaks for businesses that wanted to build in blighted urban areas. This legislation, and Hynes's determination, were responsible for attracting the Prudential Insurance Company to build Boston's first true skyscraper, the fifty-two-story Prudential Tower.

Boston was acting out of a justified sense of desperation. It seemed the right thing—the only thing—to do at the time, but in retrospect many of the initiatives and tactics of the urban-renewal era had an unintended destructive impact on the city. The New Boston was not about preserving historic places, restoring neighborhoods, or expanding recreational opportunities. Progress was defined as building new things: building an airport at the expense of an Olmsted park and a quiet residential community; building high-rise apartment buildings on the wreckage of the historic West End; building an expressway along the Charles River Basin and an interchange at Charlesgate that destroyed Olmsted's elegant entrance to the Muddy River and the Fens; and building a massive elevated highway through the city's downtown—the Central Artery.

The Central Artery was included in the critical 1948 Master Highway Plan for the Boston Metropolitan Area. Hardbound in faux red leather, the Master Highway Plan as submitted to Governor Robert Bradford was a remarkable document in the scope and boldness of its vision. That vision included Route 128, the great circumferential highway that would attract the first "high tech" businesses emerging from the laboratories of MIT and the classrooms of Harvard; the Southeast Expressway; the "Belt Route" encircling Boston; and finally, through the heart of the downtown, the Central Artery.

The Central Artery was designed to speed traffic through the city, unclog city streets, and push the city into the twentieth century. State Public Works Commissioner John Volpe, who would later be elected governor and appointed (by President Richard Nixon) federal Secretary of Transportation, waxed eloquent at a 1956 Boston College Citizens Seminar, describing the planned "integrated highway network resembling a gigantic wheel—its hub the

Central Artery and the proposed Inner Belt Route." *The Boston Globe* joined the chorus of supporters, hailing the Artery as "92 tons of steel, 459,000 tons of concrete and the widest vehicular tunnel in the world to smash the city's bottleneck."

Long considered the only solution to unclogging downtown Boston's hopelessly congested streets, the Central Artery was conceived as the hub of a new, grand regional highway system. It was, instead, a stunning example of progress run amok. Construction of the Central Artery began in 1950 and required the demolition of scores of buildings through the old downtown core, separating the North End from the rest of the city. Over 100 residences and 900 businesses fell to the wrecking ball.

State and city planners and civic leaders may have thought the Central Artery an important and essential improvement, but local residents had a different view. As construction began in the North End, the store owners and residents of the neighborhood organized to oppose the Artery. Their opposition had no effect on the relentless drive to push the Artery forward, but the clearly ugly and destructive nature of the elevated Artery structure led the builders, in 1954, to depress the portion of the project cutting through the old Leather District and Chinatown.

Building the Central Artery, a gash through downtown Boston

By then it was too late: Boston was—it seemed permanently—divided by a great, ugly steel wall that was meant to carry traffic swiftly and safely through the downtown. Lewis Mumford wrote in his book *The City in History* that the "most popular and effective means of destroying a city is the introduction of multi-lane expressways, especially elevated ones, into the Central core." Boston, Mumford observed, was a "pitiable victim [of the Central Artery] because it had more to lose, since it boasts a valuable historic core."

The Artery, splitting the city apart and damaging its historic core, did not even serve its intended purpose of easing traffic congestion. It was poorly designed with thirty-four on and off ramps—ramps that collectively took up more mileage than the Artery itself and created the unintended effect of forcing more traffic onto the elevated highway than it could bear. To top off the bad effects of the Artery, within months of its completion in June 1959, the Old Colony Line—the prime commuter rail line bringing South Shore commuters to and from the city—collapsed financially and was shut down as riders abandoned the train for the perceived pleasures of the new highway system. As more and more cars traveled on its elevated roadway, and the traffic gradually slowed to a painful crawl during rush hours, it became clear that the Artery was a complete transportation failure, a premature and expensive white elephant.

Citizens Revolt: The Tide of "Progress" Turns

BY THE 1960S some Bostonians began to understand how destructive this form of "progress" was to the identity and livability of the city. Slowly but surely, activists began to emerge in the neighborhoods—a potent combination of young rabble-rousers, grandmothers, and middle-aged professionals who, each in their own way, cared deeply about preserving the quality of life in their neighborhoods. In East Boston, when the operators of Logan Airport decided to expand the airport yet again, the Maverick Street mothers took to the streets with their baby carriages and blocked the heavy trucks that threatened to destroy their neighborhood. In the South End, activists Chuck Turner and Brad Yoneoka took a stand against the plan to run Interstate 95 through the inner city. A variety of people—from East Boston's Mary Ellen Welch, a schoolteacher turned activist who gave her life to community causes, to Milton's Elizabeth Houghton (the activists' "secret weapon" who was on a first-name basis with state political leaders like Elliott Richardson and Francis Sargent and could get their ears when necessary), to Cambridge's Father Paul McManus and Jamaica Plain's Father Thomas Corrigan, to Brookline's Guy Rosmarin—took center stage as leaders of a "people before highways" movement.

In this atmosphere of community activism, one man emerged as a smart anti-highway advocate who believed that cities were only as strong as their people—and that people would continue to abandon the city unless it offered them a higher quality of life. Frederick Salvucci was a young civil engineer in 1966 when he coauthored an Urban Planning Aid report on a transportation plan for Boston. Salvucci wrote that "Highways . . . destroy something far more difficult to replace: established neighborhoods that enjoy a deep sense of community." Even in 1966, Salvucci held a particular contempt for the Central Artery as he wrote: "Boston's Central Artery, where it borders on the North End, is a classic example of what highway critics call a 'Chinese Wall': a physical and psychological barrier isolating a neighborhood from shopping, jobs, churches, schools and friends."

Salvucci was thinking about urban transportation issues in a new way. The son of an immigrant bricklayer, Salvucci was an MIT graduate and Fulbright scholar with a talent for strategic thinking. Salvucci got to witness firsthand the lasting negative effects of poor transportation planning on a community during his years as director of the East Boston Little City Hall. Separated from the city's downtown by Boston Harbor, East Boston was nonetheless inextricably linked to the downtown because of the shortsighted decision to locate the region's major airport there. The location of Logan Airport in East Boston challenged city planners to find ways to connect the airport with the downtown and the region. In the past, those planners demonstrated a callous attitude to the residents of East Boston, demolishing homes and businesses, ruining lives, and scarring the neighborhood in order to build the harbor tunnel crossings that primarily served airport travelers and suburban North Shore commuters.

The Massachusetts Port Authority added to East Boston's transportation woes by expanding the airport's facilities. In a demonstration of raw power, Massport bulldozers destroyed the once-beautiful Neptune Road, a tree-lined boulevard of stately triple-deckers, and turned the 200-acre Wood Island Park into an extension of a runway. Wood Island Park, designed by Frederick Law Olmsted, was East Boston's only significant recreational space. Its demise sent a signal to the residents: unless they stood in organized opposition to future transportation "improvements," their neighborhood would slowly but surely be destroyed. Anna DeFronzo, a widow, grandmother, and emerging community leader, summed up the changing mood and demeanor of the neighborhood activists: "We used to go see Ed King [Executive Director of Massport], begging, begging, begging, but it didn't do any good. . . . We went to the Legislature, begging them. It hasn't done us any good. The only way is to demonstrate. There is no other way."

The destruction of Wood Island Park was an example of transportation planning that was blind to the importance of irreplaceable public amenities, and deaf to the cries of local residents who valued them. Planners in the early 1960s seemed to have no solution to the region's transportation needs that spared the residents of the city's neighborhoods. These planners took their lesson from New York's Robert Moses, who famously "raised his stein" to those who could improve blighted neighborhoods without removing people, and to chefs who could "make omelets without breaking eggs."

Predictably, it was the poorer city neighborhoods that suffered most. In the 1960s, transportation planners had targeted the Roxbury and South End communities as the unfortunate hosts for the massive highway system that would accomplish the completion of Interstate 95. The great interstate highway system begun under the Eisenhower Administration was designed to provide the people of the entire nation with modern, safe, high-speed highway connections—connections that would keep pace with the growing needs of commerce after the Second World War. The interstate highway system would also break down the barriers between cities and suburbs, barriers that highlighted the mid-twentieth-century bias that older cities were no longer sustainable as places where young families and professionals wanted to live. This bias was formed by the prevailing opinion that the suburbs, not the cities, held the promise of the "American dream"—the detached single-family house, a garage, and a small patch of lawn. Such amenities were rare in the city, and the interstate highway system made it possible for the first time to conceive of living many miles away from one's job.

This bias influenced transportation planners, who saw gritty city neighborhoods as expendable. The 1948 Master Plan was frank about its assumptions "that the relocation of tenants is an integral part of a highway project," and that "[i]n congested areas, particularly those of substandard housing nature, consideration should be given to mass relocation of tenants in new housing projects." Of course, one person's substandard housing was another's cherished triple-decker. It was true, as a 1960 Boston College report on "Travel in the Boston

Region" observed, that urban transportation planning had "not reached a level of sophistication commensurate with the importance that transportation decisions and construction have upon the health, welfare and future of American cities. Decisions in urban transportation which will shape the future city are too frequently made on a day to day, crisis to crisis, project to project basis without adequate regard for the many long range implications."

In Boston, planners conceived of a master plan for vehicular mobility that drew an arc around the city with a spike through it. This plan, with its "Inner Belt" surrounding the city, and superhighway running right through it, would require a considerable swath of some of Boston's oldest and most historic districts to be leveled, and would physically divide the city with an impassable interstate highway. A 1962 report on the Inner Belt and Expressway system took for granted the "displacement of businesses and residents" resulting from the Inner Belt in order to locate the proposed new expressway facilities "where they best serve the major traffic desires with the highest possible degree of service." The Inner Belt project was too much highway in the wrong places, and when it came time for the state planners to put their stakes in the ground, the various impacted neighborhoods—neighborhoods in Cambridge, Boston, and the surrounding suburbs—banded together in an effective effort to stave off this ultimate destructive expression of "progress." In 1969, activists from these neighborhoods formed the Greater Boston Committee on the Transportation Crisis—the GBC—a combination of "community, politicians, [and] academicians" that soon became an effective advocacy group, armed with data and substantive, policy-based arguments supplied by Salvucci and others.

The GBC was determined to halt the Inner Belt project. Roxbury's Chuck Turner (later a Boston city councilor) told Governor Francis Sargent's Transportation Secretary, "This is your world. We know it is. We have no illusions. But we won't take it lying down." Members of the GBC mounted a strong lobbying campaign in the press and at the State House and forced the governor to retreat from his commitment to the Inner Belt project. With the activists pushing in one direction and Boston Mayor Kevin White—who in 1970 was running for governor against Sargent—pushing in another, Sargent understood that he must give ground. In a televised address on February 11, 1970, Sargent put a halt to the Inner Belt. A contrite governor told his viewers, "Four years ago, I was the commissioner of the Department of Public Works—our road building agency. Then, nearly everyone was sure highways were the only answer to transportation problems for years to come. We were wrong."

The activists hailed Governor Sargent's moratorium on highway construction as a historic moment. It created an important opportunity to reconsider approaches to transportation planning. Sargent was moving in a new direction with his new policy—a momentous shift from decades of conventional thinking about transportation—and neither he nor his advisors fully understood where it would lead. State Transportation Secretary Alan Altshuler began to focus more attention on mass transit projects and on the consideration of innovative ideas like depressing the elevated portion of the Central Artery and building a third harbor tunnel to Logan Airport that would not cut through the neighborhood streets of East Boston. Although his proposed alignment—crossing the harbor diagonally from the Fort Point Channel—was later abandoned in favor of a different route, Altshuler was responsible for shifting the policy focus to transportation improvements that were congenial to urban neighborhoods.

Sargent won reelection in 1970 but was defeated for reelection four years later by former State Representative Michael Dukakis, a young and progressive reformer whose confident slogan was "Mike Dukakis should be governor." Dukakis was a fervent mass transit enthusiast, his heart and soul in synch with the anti-highway fervor of the urban activists. He appointed Fred Salvucci as his new state Transportation Secretary. Salvucci's exposure to the negative impacts of inconsiderate transportation planning on Boston's neighborhoods and citizens had left a lasting impression on him. It came as no surprise to those who knew Salvucci that, when he took office as Secretary in 1975, he set his sights on finding ways to improve public transportation and to protect East Boston from further airport-driven transportation "improvements." What did surprise them—and possibly surprised Salvucci himself—was his equally fervent advocacy for what would become the largest highway-expansion project in the city's history.

CHAPTER TWO

Laying the Groundwork: Salvucci Takes, Loses, and Regains Control

MICHAEL DUKAKIS came to office in 1975 as a reform-minded governor, challenging old assumptions and shelving existing plans as he tried to change state government's direction. In the transportation arena, one of the leaders of the "people before highways" movement was now in charge. Fred Salvucci strongly believed that Boston's urban core needed a major reinvestment in public transportation, not highways, in order to best serve the neighborhoods and revitalize the city.

The high profile of the Central Artery/Tunnel Project has obscured two of Salvucci's greatest accomplishments—the reconstruction and expansion of the Red Line, extending it north from Harvard Square and south to Quincy and Braintree (and in the process taking some South Shore commuters off the overburdened Southeast Expressway), and the rebuilding of the Orange Line (dismantling the elevated rail line that cast its dark shadow over Washington Street) and creation of Boston's most important urban amenity of the second half of twentieth century, the Southwest Corridor Park. If Fred Salvucci was committed to a coherent urban transportation planning idea, it was an idea rooted in the importance of a modern, clean, safe, and efficient public transportation system. Dukakis embodied this same view, taking the MBTA to work every day. How, then, did these two men, so anti-highway in their public-policy orientation, come to conceive, support, and lay the groundwork for the most massive highway project in the nation's history?

Fred Salvucci hated what the Central Artery represented. He had spoken out against it as early as 1966. He understood the destructive impacts that poorly planned highways and tunnels had on stable urban neighborhoods. When he took control of the state's transportation department, Salvucci was confronted with the inescapable reality that the existing Central Artery, though relatively new, had not provided the kind of transportation relief it had promised. The Central Artery was a failure, and its congested condition threatened to reach a point where highway traffic would be at a literal standstill. Built to comfortably accommodate 75,000 vehicles each day, the Artery was groaning under the weight of more than double that amount, and the volume was increasing at an alarming rate. Traffic slowed to a crawl for ten hours each day, and planners expected that by the year 2010 the Artery would be intolerably congested for sixteen hours every day. The congestion hurt the economy, damaged the environment, and marked Boston as a place to be avoided during nearly all times of the day. The ever-expanding demand for access to and from Logan Airport put additional pressure on the overburdened urban transportation network. The business community joined the Massport and MassPike planners who demanded action. Their unimaginative solution was to build yet another harbor tunnel from downtown into East Boston.

Although he opposed highway expansion, Salvucci could not ignore the need to respond to these growing pressures. The question was: how to solve these problems in the most progressive, people-friendly, neighborhood-friendly way? Could the city be improved by highway expansion, or would the past repeat itself as more and more lanes of traffic destroyed what was left of the integrity of the downtown and its neighborhoods?

The Big Dig: The Early Strategy

SALVUCCI, WHO IN LATER YEARS modestly described his role as an "assembler" of ideas, credits another MIT engineer, Bill Reynolds, as the inspiration for his thinking about the Big Dig. Salvucci recalls that Reynolds approached him with the idea of putting the Central Artery underground. Reynolds was no community activist. He was a pro-highway contractor who saw the Central Artery as a problem for highway advocates because it was "a giant billboard

Fred Salvucci, father of the Big Dig

that says roads are bad." Reynolds thought that depressing the Artery would eliminate this poster child for the anti-highway advocates.

Salvucci initially dismissed the idea. "I thought [Reynolds] was crazy," he recalled. "What are we going to do, put up a sign at the Charles River saying 'City Closed For Alterations. Come Back In Ten Years'?" Salvucci could not imagine how such a massive construction effort would take place in a city that was so congested, so old, and so dependent on what little mobility it had. How could you build an underground interstate highway system without disrupting the existing elevated highway? The challenges appeared insurmountable. But Reynolds persisted. "This is a solvable problem," he told Salvucci. And as Salvucci thought more about it, he could see the possibilities—he could see a strategy unfolding.

For years, pressure from the business community to build a third harbor tunnel to serve the airport threatened the East Boston neighborhood. Plans drawn up by the Turnpike Authority and others had the proposed new tunnel emerging right in the middle of the community, cutting off and isolating the historic and tightly knit Jeffries Point neighborhood. The third harbor crossing envisioned by the Turnpike planners would be the death knell for East Boston. But Salvucci was thinking outside the box. What if you could figure out a way to improve mobility and not destroy neighborhoods? What if you could improve highway traffic through the city, and increase capacity to the airport, and actually improve the quality of life in the neighborhoods while doing it? In theory, you could have your transportation cake and eat it too if you could widen and submerge the elevated Central Artery, and if you could bring a new tunnel to the airport on an alignment that placed it entirely within airport property. The business community and Massport wanted a new tunnel badly—they might bargain for a comprehensive solution to the city's transportation problems. And so

Salvucci began to devise a strategy that linked the Artery improvement and third harbor tunnel initiatives. "I knew the politics would be better if we did a joint project," recalled Salvucci. "I knew that in my stomach."

Linking the depression of the Central Artery with the construction of a third harbor tunnel was a masterstroke that enabled Salvucci to forge an alliance between the business community and the neighborhoods. Salvucci at this time also considered the potential opportunity of including a link between the city's unconnected north and south rail lines. The north-south rail connector was an idea that charmed many pro-transit enthusiasts, including Dukakis. But given the limited amount of space under the proposed new Artery alignment, and the additional cost and fairly limited benefits of a north-south rail connector, building it was never a realistic possibility.

Salvucci began to plan this bold project with one unchangeable principle: not one home would be displaced. For once, a major highway initiative would not come at the expense of the quality of life in the neighborhoods. Indeed, it would improve those communities by opening up opportunities to re-knit neighborhoods and create significant open space and recreational amenities for the entire city. "I saw," recalled Salvucci, "that it was now appropriate to look at interstate highways in a different way—through the lens of environmental impact, urban design, even socioeconomic justice."

The first thing Salvucci did was secure a commitment of federal funding for his idea. Without a sufficient federal funding commitment, the Artery/Tunnel idea would never become reality. It would be far too expensive for the state to support on its own. But once the federal government committed to a state interstate project, federal highway funds would support 90 percent of its costs.

Salvucci had an important insight: that the 90 percent federal interstate completion funds that were potentially available to Massachusetts could only be used for a project like the Big Dig. The state could not use these funds for public transportation or other state highway initiatives.

The Federal Highway Administration (FHWA) decided how much money states would receive for interstate highway projects. Each state would submit an interstate completion estimate (ICE) of the costs of congressionally authorized interstate completion projects in that state. Once Congress authorized a transportation project and included it in a state's ICE, there was a high likelihood that the funds would be available when needed to actually build the project. Federal rules required funding until completion—meaning a potentially unlimited federal financial contribution to the Artery project.

But the FHWA bureaucrats were angry at Massachusetts political leaders for abandoning the Inner Belt. They reacted skeptically to Salvucci's request for $360 million for the depression of the Central Artery. The FHWA also argued that the depression of the Artery appeared to offer limited or no transportation benefits. This became a constant refrain from the federal government. Salvucci made full and effective use of the state's powerful congressional delegation, led by House Speaker Tip O'Neill, to persuade FHWA of the merits of the state's request. With pressure coming from the House Public Works Committee, FHWA relented and included the requested amount in the ICE.

In the final approvals for the Artery/Tunnel Project, the federal government would not allow interstate highway funds to be used on the portion of the Project from High Street to Causeway Street. This decision, which would come later in the process, did not trouble Salvucci—he did not view the High to Causeway portion as a difficult or expensive element of the Project. For now, Salvucci had protected his position with a clear federal commitment—an initial one to be sure, but an important foot in the door for Massachusetts.

The King Years: Salvucci on the Sidelines

SALVUCCI WOULD HAVE PRESSED ON with his ambitious Central Artery/Tunnel Project except for a four-year interruption of his plans. Governor Dukakis's unexpected defeat for re-nomination by former Massport Executive Director Edward J. King in 1978 was Fred Salvucci's worst nightmare. King had literally and figuratively bulldozed his way into power at Massport. He was a bottom-line businessman, a person of extraordinary single-minded focus on achieving his goals. King brought Massport into the jet age, expanding its facilities into a large and successful international airport, but he did so at the expense of the East Boston neighborhood and without a vision of what would happen when the airport reached capacity. When Governor Sargent reversed course and established a moratorium on highway growth, King pushed on with his airport-expansion schemes, including a plan to build a third harbor tunnel into East Boston. His relationship with Sargent grew strained, and in the months just before Dukakis took office in 1975, the Massport Board—appointed by Sargent—voted to fire King.

King watched from the political sidelines as Dukakis and Salvucci implemented their pro-mass-transit policies. Most political observers wrote

King off as a gadfly when he announced his intention to run against Dukakis in the Democratic primary. But Dukakis had taken office during difficult economic times, and had been seriously hurt by both a tax increase early in his term and severe cuts his administration was forced to make in human-service programs, cuts that alienated his natural constituencies. King's stunning victory in the Democratic primary and his general-election victory over Republican Francis Hatch meant that Dukakis and Salvucci were out, and the new governor would be the man who had expanded Logan Airport by destroying Wood Island Park, and whose trucks had been blocked by the Maverick Street mothers. The state's transportation policy debate appeared to have come full circle.

KING'S VICTORY REVIVED THE EAST BOSTON community activists. They knew that a third harbor tunnel was one of the new governor's transportation priorities. This was, indeed, the case. King thought a new harbor tunnel essential to regional access to Logan Airport, and he proposed aligning the tunnel's East Boston portals right up the old Conrail corridor, splitting the community in two, and demolishing several homes and businesses. This same alignment had been recommended by the Turnpike Authority to Governor John Volpe in a June 1968 report declaring the "inescapable" conclusion "that a Third Crossing of Boston Harbor is urgently needed and needed immediately." The Turnpike Authority's 1968 report declared that after studying "all practical tunnel alignments," it was "determined that no feasible location" could be found that did not take East Boston residences. Knowing this, East Boston citizen activists formed the Coalition Against the Third Harbor Tunnel—CATT—to organize opposition to King's plans. Fortunately for these activists, the King Administration could never organize itself well enough to make the third harbor tunnel a serious threat.

Ed King's four years as governor were marred by one misstep after another. Two of his senior-level appointees admitted to lying on their résumés. King's Secretary of Transportation was sent to jail after being caught red-handed taking a cash bribe. And a scandal at the Department of Revenue led to the suicide of one of King's Deputy Tax Commissioners. In this environment of near constant turmoil, King had a difficult time establishing a firm grip on the government's policy-making apparatus. In particular, the disruption caused by scandal and mismanagement in his transportation secretariat, and delays caused by middle managers remaining from the prior administration who did not support King's policies, meant that King never was able to change the policy direction that had been established by Sargent and Dukakis.

After his Transportation Secretary went to jail, King appointed successful businessman James Carlin to clean up the mess of corruption left behind by his convicted predecessor. Carlin embarked upon a forceful effort to regain the initiative, making no secret of his disdain for the Dukakis/Salvucci brand of transportation policymaking. In an op-ed piece in the *Boston Globe*, Carlin asked: "What has Massachusetts started and finished in terms of major road improvements inside Route 128 since 1960? The answer is almost nothing. . . . Public officials must understand that projects must be built and that vocal community groups must be carefully heard but not allowed to dominate the greater public good."

Carlin might have had the stamina and skill to push the third harbor tunnel ahead, as he hoped to do, but he never had the chance. In a 1982 primary rematch Dukakis defeated King, and Carlin and his team returned to the private sector. One of Carlin's younger protégés was James J. Kerasiotes, a talented, tough, and brash public manager with ambitions to restore his mentor's "can do" attitude to state government. Kerasiotes left office taking with him an important political lesson: if you were going to run a government agency effectively, and get results, you needed to take control quickly over the people running the agency. If you let holdovers continue in office, and if you did not immediately articulate a clear set of policy directives, your policy agenda would never get off the ground. Kerasiotes had eight years under the next two Dukakis terms to reflect on this lesson.

CHAPTER THREE

Planning for the Big Dig: Politics, Persistence, and Perseverance

IN 1982 MICHAEL DUKAKIS reclaimed his hold on the state Democratic Party and the electorate, easily defeating King in the Democratic primary and winning the general election against a weak Republican candidate. Dukakis remained in office as governor for the next eight years. Through all that time his Transportation Secretary was Fred Salvucci.

Dukakis ran for reelection affirming his pro-mass-transit platform, and he made common cause with the East Boston activists who opposed King's third harbor tunnel plan. Dukakis was "dead-set against the [third harbor] tunnel" because the "King plan was just a disaster. It came up right in the middle of a neighborhood."

Salvucci understood where his boss was on transportation issues. He also knew that the business community was solidly behind a new tunnel to the airport. Salvucci needed Dukakis to support his vision for a combined Artery/Tunnel project, even though that meant increasing interstate highway capacity, and he needed the business community to support the Artery depression, even though his solution to urban traffic congestion seemed too expensive, too complicated, and potentially too disruptive of the city to be realistic.

Salvucci's solution was to forge political and community support for both a new airport connection that completely avoided the East Boston neighborhood, and a depressed Central Artery that would significantly increase capacity and offer the city a once-in-a-lifetime opportunity to open up its old downtown core. Dukakis's Economic Development Secretary, Alden Raine, described Salvucci's "great vision" as identifying the only alignment for the inevitable third harbor tunnel that would both accommodate airport traffic growth and protect the quality of life in East Boston. Salvucci's years of work on behalf of the East Boston community gave him a large measure of credibility that enabled him to move the idea forward. Joseph Aiello, a community activist who would later become a transit specialist in Salvucci's office, recalled that the community believed that "Fred would get the nuances right." In an environment usually marked by distrust, Salvucci had earned the initial confidence of the neighborhood residents.

Back in office as state Transportation Secretary in January 1983, Salvucci faced a significant initial hurdle: a federal deadline for filing an environmental impact statement, or EIS, by September 1983. He faced another and possibly even more difficult barrier: while Dukakis and Salvucci had been out of office,

The Central Artery. Boston's failed experiment became a parking lot during rush hour.

Ronald Reagan had been elected President. Reagan had praised Ed King as his "favorite Democratic governor." Now that Dukakis had defeated King, it was difficult to imagine Reagan's administration being favorably disposed toward Mike Dukakis and Massachusetts.

The new Federal Highway Administrator, a highly partisan Texan named Ray Barnhart, brought a visceral dislike of Massachusetts to his decision making. Barnhart, who would eventually come to support the Project at a crucial moment, was initially "ticked that Massachusetts had screwed around for so many years in such an anti-transportation mode with all their social maneuvering. Coming from Texas, my first inclination was to let the bastards freeze in the dark." Fortunately for Massachusetts, Barnhart and his deputy Les Lamm decided not to make Salvucci's challenge any more difficult than it already was. Lamm recalled that he and Barnhart "could have played hardball [but we] decided to give them the opportunity to keep the project alive."

Salvucci developed a federal approval strategy centered on two men: Barnhart and Tip O'Neill, the powerful Speaker of the House whose district covered a large portion of the Project area. Barnhart would be essential to the Project's final approval, but Salvucci understood that Massachusetts would not stand a chance at the series of federal approvals that were required without the strong and unwavering support of the man who lived and died politically by one simple maxim: all politics is local.

Tip O'Neill to the Rescue

IN HIS ENTIRE CAREER IN CONGRESS, Tip O'Neill's only significant local political opponents were candidates from East Boston. Another man might have made an effort to remove that politically charged neighborhood from his district, but not O'Neill. Secure in his Cambridge base, he tried to support and understand the particular political needs of East Boston and became that neighborhood's champion. In the 1960s and 1970s, that meant being a staunch opponent of Massport and the expansion of East Boston's already crowded transportation infrastructure. O'Neill understood local retail politics as well or better than most, and he was very careful to never be caught on the wrong side of a local district issue.

When Salvucci brought his Artery/Tunnel plan to O'Neill, the Speaker was taken aback. "What's this?" O'Neill asked Salvucci, as the Transportation

Secretary rolled out a map of the proposed project. When Salvucci explained that the Project included a new third harbor tunnel, O'Neill barked: "What tunnel? We're not building any tunnel." Congressman Joseph Moakley, representing South Boston and intimately familiar with the city's political dynamic, recalled that "Tip and I were not too crazy about the third harbor tunnel, because Tip felt that his people in East Boston would be dispossessed and I felt my people in South Boston would be removed." But Salvucci pressed on and made his case.

Congressman Brian Donnelly of Dorchester recalled that O'Neill was "a little cold to Fred. He's got this big cigar and the ashes keep flipping on the plans as he says 'What about so-and-so?' He keeps throwing out names. 'Are the people in Eastie for this?' And I'm convinced the whole thing, literally, is going to go up in flames. But Salvucci is right there with him, saying 'Yeah. Yeah. They're for it.'" Moakley recalled Salvucci saying, "Just give me a chance, and if at any stage of the game, before the bill takes effect, either one [of you] says 'Stop', that's the end of the Third Harbor Tunnel." O'Neill later confirmed that his "only question" was whether East Boston would be harmed by the Project. When he became convinced that East Boston was safe, O'Neill recalled, "I didn't have any difficulty. I have to take care of my locale first."

O'Neill's hand-picked chairman of the House Public Works Committee was a New Jersey Democrat from a traditionally Republican seat, Jim Howard. O'Neill was very close to Howard and helpful to his election to the House, and Howard in return was thoroughly loyal to O'Neill. "Howard idolized Tip. Whatever Tip wants, Howard is for," recalled Brian Donnelly. With O'Neill and Howard firmly on his side, Salvucci could turn his attention to the Republican side of the aisle. He was convinced that he could persuade congressional and administration Republicans that they, too, should support this massive federally funded project for Boston, Massachusetts. Salvucci, a vegetarian who eats neither meat nor fish and spurns leather clothing, understood what drives politics: "There are no vegetarians in Congress when it comes to pork."

Securing Republican support would not be easy. Senate Majority Leader Bob Dole, whose wife, Elizabeth, was Reagan's Transportation Secretary, identified the Project as an example of "parochial special interests" that were causing federal deficits and had "no real purpose in our national transportation scheme." Salvucci determined that in order to protect his Republican flank, he needed the advice and active support of nationally recognized Republicans.

He found a key supporter in Boston lawyer Roger Moore, who had strong Republican credentials going back to the Goldwater campaign of 1964.

By choosing the Bechtel Corporation (joined with a joint venture partner, Parsons Brinckerhoff) to be at his side as the Project's overall manager, Salvucci brought the cachet of a nationally respected firm that had close connections to Reagan cabinet secretaries Caspar Weinberger and George Schultz. Salvucci's decision to privatize the management of the Project had deep roots in his own distrust of the state's Highway Department and his well-considered view that the department was not up to the task of managing such a complex and massive project. This decision also enabled the Highway Department to focus on delivering transportation improvements to the rest of the state—an important political consideration since it would be disastrous if the majority of the state's legislators believed that their local needs were being neglected at the expense of the "Boston project."

O'Neill wanted the Project approved by the Federal Highway Administration and he wanted the Project included in the federal ICE authorization. Getting the Project included in the ICE was an essential step from which there would be no turning back. A congressional staffer explained that "getting a project included in the Interstate Cost Estimate was like gold. Once it was in the Interstate Cost Estimate, it meant there was an obligation [on the part of the federal government] to provide the states the money to finish it." But Republicans controlled the White House and federal bureaucracy, and from 1981 through 1986 they also controlled the Senate. Getting House approval was one important step—but not the final word on whether Massachusetts would receive approval for its ambitious Artery/Tunnel Project.

Thus began Salvucci's long struggle to secure federal approval of the Project. His chief aide helping him navigate the perilous waters of the federal capital was Charles (Chip) DeWitt, an affable lawyer with a quick mind and charming manner who would spend weeks at a time holed up in a Washington hotel room supervising the Washington lobbying effort led by John Cahill. Through O'Neill and Howard, Salvucci was able to keep federal approval for the Artery/Tunnel Project on the table as a Democratic Party requirement in every transportation bill discussion during the mid-1980s. This display of power did not go unnoticed by Washington professionals. The need for compromise was clear, and Salvucci's Republican consultants made FHWA Administrator Barnhart their principal target. Salvucci and his staff met with Barnhart time and again, in Washington and in locations across the country—wherever Barnhart would be at a conference or meeting of transportation officials—and they pushed hard to understand what the irascible Texan would consider adequate for FHWA approval.

Salvucci and Barnhart finally reached an agreement: if Massachusetts could demonstrate that the benefits of the depressed Artery were equal to the benefits of the third harbor tunnel—a tunnel that Barnhart liked—then FHWA would approve the entire deal. The two men agreed to have two teams—a legal team and an engineering team—wrestle the outstanding issues to the ground. Their staffs battled over arcane legal issues and cost/benefit analyses. Dozens of alternatives were explored, and comparisons made, to demonstrate the beneficial effects of the proposed Project.

Matt Coogan, Salvucci's principal aide in developing the initial overall plan for the Project, worked closely with FHWA staff in a strenuous back-and-forth exchange of data, statistics, and forecasts in order to demonstrate that the entire Project as Salvucci envisaged it did, indeed, provide significant transportation benefits that were worthy of federal support. Coogan would respond to FHWA questions, only to be given new questions to answer, new alternatives to evaluate. This he did for months, and in a game of patience and persistence the Massachusetts officials outlasted their FHWA counterparts. In the end, Barnhart was persuaded. In a 1985 memorandum, Barnhart tipped his hat to the "positive although limited transportation benefits" of the Project, as well as "the State's prerogatives in the use of federal-aid funds." While his support was lukewarm, Barnhart took a "hands-off attitude" toward the Project. "Will I endorse the Project? No. Will I oppose the Project in Congress? I probably will not."

Barnhart's decision to keep "hands off" the Project was made in a highly charged political environment. With FHWA no longer in opposition, the Democrats in the House saw their opportunity to finally lock in federal funding for the Project. O'Neill was not running for reelection in 1986. He knew that he could rely on Jim Howard and his successor, Speaker Jim Wright, to stay the course for their friend and mentor from Massachusetts. And in the general election, the Democrats returned to power in the Senate. With New Englander George Mitchell in Senate leadership, Salvucci was optimistic that in 1987 he would finally get his federal approval. There was just one more person standing in the way of success—the President of the United States.

Struggle in the Senate: Massachusetts Rises to the Occasion

RONALD REAGAN NEVER LIKED the Central Artery/Tunnel Project. Nor did Reagan's Transportation Secretary, Elizabeth Dole. Reagan was also opposed to increased federal spending across the board. When Congress, in 1987, finally passed an omnibus transportation bill that included funding for the Artery/Tunnel Project, Reagan's Office of Management and Budget advised him to veto it. The bill, in their view, contained too many special interest projects—and the Big Dig was high on that list. Reagan vetoed the bill on March 27, 1987, singling out the Project as particularly excessive and emblematic of the "unjustifiable funding for narrow, individual special interest highway and transit construction projects." Reagan knew a good issue when he saw one, and lambasted the bill with gusto. "I haven't seen this much lard," said the affable President, "since I handed out blue ribbons at the Iowa State Fair."

But the Reagan veto was no joke to Massachusetts. Overriding the veto became a high stakes political game. Now Senator Edward M. Kennedy led the entire Massachusetts delegation as it worked in overdrive to find the votes to override the President's veto. Their job in the House was easy enough because of the huge Democratic majority in that body. But in the Senate, the vote was viewed as perilously close. On April 1, 1987, thirteen Senate Republicans—Republicans willing to abandon Reagan because of the strong lure of transportation projects in their own districts—voted with their Democratic colleagues to override the veto. Every Democrat senator voted to override the veto except one: the senator from North Carolina, Terry Sanford.

Sanford had promised his fellow North Carolina Democrat, Lieutenant Governor Bob Jordan, that he would vote to sustain Reagan's veto in order to help Jordan in an upcoming election against a Republican congressman who had opposed the bill. Caught between this promise and loyalty to the Democratic Party in Washington, Sanford initially voted "present," and then changed his vote to "no," leaving the final tally one vote short of the necessary two-thirds. During what the *Boston Globe* described as a "tension-filled roll call," Sanford "agonized for an extraordinary ten minutes in full view of his colleagues and public spectators." Unless he changed his vote the veto was sustained. There would be no Big Dig.

A political drama unfolded, the likes of which had rarely been witnessed in the Senate. Everything was done to persuade Sanford to switch his vote: Ted

Kennedy appealed to Sanford's old ties to the Kennedy family, Majority Leader Robert Byrd offered instruction in the importance of loyalty, others threatened withdrawal of New England support for North Carolina tobacco subsidies. Lieutenant Governor Jordan of North Carolina was warned of retribution by his state's congressional delegation if he did not release Sanford from his commitment.

The pressure on Sanford was, in the characterization of one staff member, "brutal." The drama played itself out over several hours, and for Sanford the choices were about as bleak as they get: if he held to his vote, he would risk his position as a loyal party man, and he would likely never again get an accommodation for his state on any matter, large or small. If he switched his vote, he would be remembered as a man who could buckle under pressure, whose promises did not matter. It was an unenviable position to be in.

Sanford finally announced that he was changing his mind, and his vote. A few tense days lay ahead as Reagan turned on his famous political charm in an effort to pry away a vote from the override column, but it was to no avail. On the day of the final vote, Byrd and Kennedy served as unofficial chaperones for Sanford—just in case. When Majority Leader Byrd brought the matter up for reconsideration, the veto was overridden by one vote. One Senate staffer said, "I've never seen anything like it before or since."

Salvucci and his staff celebrated having "foiled the skeptics in this city," and Senate President William Bulger declared that the Project would allow "every man, woman and child in Massachusetts to work in construction for the next decade."

The transportation spending bill was now law. But the long battle for federal funding exacted a significant cost. The popular impression was that Massachusetts, by winning the veto override, had secured 90 percent federal funding. But that was not the case. Since funding approval did not come until 1987, and since it would take several more years after that to complete the work necessary to achieve key federal and state environmental approvals, the Project faced the stark reality that work would begin at the tail end of the interstate highway completion process, and there was only one federal transportation bill remaining to fund these interstate completion projects.

Salvucci would remind anyone who would listen that Massachusetts had historically come out on the short end of federal financial support for its transportation infrastructure. "The reality is that the original Central Artery, the Tobin Bridge, the Sumner and Callahan Tunnels, the entirety of Interstate 90 [the Turnpike] and Route 128 all were built without one penny of federal funding." Now, when Massachusetts finally presented a coherent solution to its highway congestion problems, the delay in federal funding approval meant that Massachusetts would not benefit from the same federal largesse that every other state had enjoyed.

"The fact that Reagan and his people succeeded in delaying Project approval, and therefore construction until after 1991 meant two things," Salvucci pointed out. "First, inflation would cause costs to rise, and second, it pushed the Project out of the 90 percent federal funding era and into the ISTEA era where national interstate spending was capped at six billion dollars. So the idea that Massachusetts rolled the federal government is simply not true." To the contrary, says Salvucci, even with the considerable federal support for the Big Dig, Massachusetts had never received its fair share of federal funding for its interstate highway system.

The delays would be harmful to the bottom line, but the Project was finally approved and could now move forward. The *Boston Herald* featured a front page photograph of a clogged Central Artery with the headline "Relief!" printed over it. It was the end of an important preliminary phase. Now, the Project had to be built.

CHAPTER FOUR

The Homefront: Salvucci Cobbles Together the Local Coalition

DURING THE YEARS Salvucci was focused on gaining federal support, he was also busy at home, pulling together a complex and sometimes fragile coalition of community and business leaders to support the Project. Salvucci was a master tactician, a man described by Boston Mayor Kevin White as having "great quality, strength and tenacity." White observed that these traits did not develop "overnight," adding that Salvucci "has honed it to the point where it bears on the awesome." Salvucci's Communications Director, Michael Shea, recalled that "if five people got together in the North End, Fred would be there to brief them on the Project." Night after night, Salvucci would meet with local residents and politicos, sometimes one on one, sometimes in large groups, always determined to persuade people that the Project would have a positive influence on their neighborhoods, that it would improve their quality of life, that they needed to give him this chance and believe in him. It was not always easy. Highway planners in the past had lied and bullied their way through the destruction of neighborhoods when building new highway systems. Salvucci, who understood that "dialogue is not a substitute for action," was asking these neighborhood activists to forget the past and believe that he could introduce a truly new way of doing business as he constructed the most massive highway expansion project in the city's history.

Out of the ashes. Spectacle Island, transformed from garbage dump to public amenity, has been reclaimed.

Salvucci never moved from his one nonnegotiable requirement that not one home or active downtown business would be taken for the Project. While this commitment did not carry over to businesses in the South Boston industrial seaport area, or to the Park & Fly business in East Boston, it applied with unrelenting focus in the downtown. Salvucci also gave neighborhood activists the ability to influence the discussion about what the new parks and open-space features coming out of the Project would look like.

Hurdles still abounded as Salvucci worked to keep his local support intact. When the Gillette Company, one of Boston's largest private employers, raised serious questions about the potential impact of construction on their South Boston manufacturing plant, Salvucci responded with an elaborate design and construction solution, which included building one of the world's largest dry docks in order to construct the immersed tunnel tubes that would comprise the elaborate Fort Point Channel crossing. In East Boston, Salvucci had to find a way for the new access ramps into the airport to avoid the existing Delta Reservations Center, a small block of a building that operated twenty-four hours every day and employed over 500 people. And in the Financial District, Salvucci sparred with Donald Chiofaro, the brash developer of International Place, a monstrosity of a building that even its architect, Philip Johnson, later derided as too much space in the wrong place. Chiofaro defied the Project by building an 800-car garage alarmingly close to where the underground Artery would be, and then demanded the ability to take down an existing Artery ramp to build one of his project's towers. After years of negotiation, Salvucci finally agreed to let Chiofaro complete his building, with a relocated ramp built at Chiofaro's expense and a ten-foot utility corridor kept available for the Project's use.

Salvucci also made common cause with the city's environmental community. Unlike every other transportation expansion project in the city's memory, this Project as Salvucci explained it would actually improve both mobility in and through the city and the quality of life in the neighborhoods. It became increasingly clear that when completed, the Project would lead to the greening of the city in ways even the most optimistic environmentalists could not anticipate.

Mitigation: The Greening of Boston

SALVUCCI HAD MASTERED THE ART of reaching out to community leaders and empowering them to have an impact, within carefully defined limits, on design and mitigation issues. Salvucci was keeping faith with the people he had joined and supported in his early days as a fellow activist, but he also recognized the potentially destructive power of organized activists in places like East Boston and the North End. In Salvucci's mind it was the right thing to do, but it was also good politics. And it was necessary for the Project to pass numerous federal and state environmental law requirements.

The impressive and costly mitigation efforts, and design changes made to satisfy neighborhood activists, underscored that this Project was unlike any other highway expansion project in the city's history. Boston was getting a new interstate highway system, and a new harbor tunnel, but the Big Dig was more than a mere highway expansion project—it was a city improvement project. The Big Dig would create over 300 acres of new open space and parkland, a completely new and catalogued underground utility system, and significant public transit improvements. The most visible element of the greening of Boston would be the Rose Kennedy Greenway, a downtown expanse of passive and interactive open-space features on the footprint of the old elevated Central Artery that will reconnect several Boston neighborhoods—the North End, Chinatown, and the Waterfront—with the downtown financial and civic districts. The Greenway was a once-in-a-lifetime opportunity for a city of Boston's age, and it would be a civic improvement of incalculable proportions—its ultimate impact not unlike that of Central Park on New York City—but that was only the beginning of the story.

The Charles River Basin, the great public amenity created at the beginning of the twentieth century, now would be improved and connected to Boston's Harborwalk, offering citizens the chance to walk from the Charles River Esplanade through new parkland directly to the Boston Harbor waterfront. Forty acres of new parkland on the Boston and Cambridge sides of the river were reclaimed, new seawalls built, and landscaping installed. To the south of the downtown, at the Fort Point Channel, some 4,200 feet of seawalls were reconstructed and the Boston Harborwalk expanded by 2,500 feet.

One of the Big Dig's most important and lasting improvements may be the reclamation of Spectacle Island. Unlike many other waterfront cities, Boston has never fully made use of its harbor island system. Over the years Spectacle Island was a glaring example of neglect and environmental abuse. Once the site of a horse-rendering plant, a grease extraction plant, and finally a city garbage dump, the island was such a man-made toxic cesspool that it would spontaneously catch fire as a result of the combustion of methane gases emanating from the tons of rotting refuse. Project officials endured laborious discussions and negotiations with several federal agencies, notably the regional office of the Environmental Protection Agency, but they finally procured the approvals necessary to fill in Spectacle Island with up to 2.7 million cubic yards of fill from Project excavation. In the end, the Project made Spectacle Island a fully functioning recreation site, with two beaches, five miles of hiking trails, landscaping including 28,000 trees, shrubs, and vines, a ferry ramp system, a visitors center, and a 550-foot pier with berths for thirty-eight boats.

Two environmental improvements were particularly ironic. In Revere and Saugus, where, decades earlier, transportation planners had proposed routing the completed Interstate 95 thorough valuable wetlands, the Project restored eighteen acres of Rumney Marsh, a salt marsh that had been destroyed before Governor Sargent put a halt to the project. And in East Boston, long suffering under the weight of airport expansion and the two harbor tunnels, the Project provided fifty acres of new open space, including a twenty-acre park along Bremen Street. This was a salient example of how the Project could improve the quality of life for neighborhood residents—and the new park was built on the exact land that the King Administration and the business community had proposed for the original third harbor tunnel alignment. Now, rather than having another tunnel divide their community, East Boston residents have a greenway connecting their community to new waterfront parks. Each neighborhood impacted by the Project—Charlestown, the North End, East Boston, South Boston, and Chinatown—became the beneficiary of new open space or parkland.

The environmental and civic benefits of the Project were being enjoyed throughout the region. Numerous cities and town throughout Massachusetts, and some in Connecticut and Rhode Island, used Boston blue clay dug out from the Big Dig construction site to cap landfills or fill in dangerous abandoned quarries.

The Project did not simply expand the city's open space, it improved its built environment. Two examples stand out: the Leonard P. Zakim Bunker Hill Bridge crossing the Charles River, and the several ventilation buildings that dot the Project at various points. The Zakim Bridge took on iconic proportions as soon as it was completed—the widest cable-stayed bridge in the world, carrying ten lanes of traffic held aloft by elegant, white cables that are beautifully lit at night. It is Boston's Golden Gate, a technological and architectural marvel that helps define the energy and vitality of a city.

The various ventilation buildings are an equally important example of the sensitive manner in which the Project designed its facilities. Since most of the Project is underground, and since tunnels require regular ventilation, several ventilation buildings needed to be built along the entire Artery/Tunnel system. Prior ventilation buildings serving the Sumner and Callahan Tunnels were monolithic, Soviet-style buildings without charm or respect for their surroundings. The ventilation buildings designed for the Project were elegant examples of form following function—at the airport, topped by stainless steel caps that glisten in the sun and glow with color at sunset; at Haymarket, surrounded by a parking garage, MBTA station, and retail and office space; and in the South Boston waterfront area, clad in bright and vibrant brick and colored panels.

Fateful Decision: MassPike Gets to Run the Big Dig

DURING THE FIRST HALF of the twentieth century, one man dominated transportation planning in Massachusetts. William F. Callahan was a single-minded advocate for highway construction, a man who believed fervently that the future lay along a well-paved multilane expressway. Callahan was a savvy politician who understood power and its uses like few men of his generation. His appointment as first chairman and CEO of the Turnpike Authority in 1952 capped a career in which he served as state Highway Department Commissioner from 1933 to 1939, and again from 1949 to 1952.

Under Callahan's influence, the legislature created the Turnpike Authority as an autonomous, financially independent authority, protected from the shifting political sands of Beacon Hill. That independence was meant in part to give confidence to the investors who would purchase the Authority's bonds, and in part to insulate the state's elected leaders from the difficult tasks of raising and collecting tolls. Callahan led an agency governed by a three-person board. He, as full-time chairman and CEO, was its undisputed master. In a time when governors were elected to two-year terms, each board member had a renewable eight-year term, which meant that no governor could control the direction of the Authority or its access to hundreds of millions of dollars in annual toll revenues.

Callahan put together a talented and fiercely loyal team of engineers, lawyers, and financial advisors. Like his New York counterpart Robert Moses, he let no person or obstacle stand in his way. He built the 135-mile Massachusetts Turnpike in record time—the first 127 miles from the New York border to Route 128 was begun in 1954 and completed in 1957. Callahan also built the second harbor tunnel, which was named in honor of his son and namesake, who was killed in the Second World War.

When Callahan died in 1964, the powerful position of Turnpike Chairman was a political plum beyond all others. Governor Endicott Peabody, in a fight for his political life and needing all the help he could get from established political machines, awarded the chairmanship to State Treasurer John Driscoll. Driscoll served as Turnpike Chairman from 1964 until his retirement twenty-three years later. In that time MassPike undertook no building program and operated its transportation facilities quietly and without fanfare.

By the mid-1980s, the Turnpike Authority was generally perceived to be a dinosaur, serving no useful purpose as an independent entity. Its leadership remained a mix of aging Callahan acolytes and Driscoll relatives and allies who had comfortably settled into what they perceived as lifetime appointments. They were nearing retirement age, and they were bereft of creativity and energy. Under their watch, the harbor tunnels were left to deteriorate to a condition that came close to threatening the safety of motorists who used them.

The turnpike required significant reconstruction, and its eleven service areas were environmental time bombs, featuring antiquated single-walled underground storage tanks that were leaching gasoline throughout neighboring communities. The one piece of good news was that, because of its inactivity, the Turnpike Authority was a genuine cash cow with enormous bonding capacity.

It should not have surprised anyone that Salvucci looked to the two great independent authorities—the Massachusetts Port Authority and the Turnpike Authority—as important instruments in his overall Artery/Tunnel plan. These authorities were created in the 1950s by a legislature concerned that the state could not, on the strength of tax revenues alone, build and maintain the important elements of a modern transportation infrastructure, namely, the international airport and the east/west interstate highway. It was important that these authorities be able to raise revenues independently from state government. The need to establish their financial independence to the satisfaction of investors meant that the legislature granted them a large measure of political and policymaking autonomy. The most significant guarantee of that autonomy, which was not lost on Salvucci as he planned for the future, was the appointment of Massport and MassPike board members to staggered terms. Salvucci knew that, barring resignation or death of a board member, the new governor following Dukakis would not get control over Massport until 1995, and MassPike would remain out of reach until 1996.

Salvucci initially thought that Massport would have primary responsibility for the Project. This appeared to make sense; Massport historically was a well-managed agency and as an independent revenue bond authority had access to a variety of significant revenue streams. Salvucci was also always looking for ways to spend Massport's money—it was his way of crippling an agency he did not trust, so that its constant urge to expand would be constrained by limited revenues. Further, Massport would directly benefit from the third harbor tunnel, so it made sense for it to oversee the Central Artery/Tunnel Project.

But Massport recoiled from taking on a significant role in the Artery/Tunnel Project. Its leadership rightly viewed its central missions as aviation and economic development. Massport was also legendary for being an agency beyond political control. Even when Dukakis and Salvucci had appointed the entire Massport board, the Port Authority's staff exercised stubborn independence and failed to follow Salvucci's directives to the letter. Moreover, Massport had its own ambitious improvement program that would eat up a billion dollars of its financial resources.

Salvucci also began to focus increasingly on the post-construction costs of the Artery/Tunnel Project. While most eyes were on design and construction costs, Salvucci directed his consultants Lazard Frères to estimate the costs of maintaining and operating the miles of tunnels under the city and the harbor. Their answer: $60 million a year, an estimate Salvucci believed was conservative. Where would those funds come from? The need to allocate the operation and maintenance burden fairly led Salvucci to consider sharing the burden between Massport and the Turnpike Authority. He and his senior staff began to consider what it would take to prepare the Turnpike Authority to assume the role of owner and operator of all or a significant portion of the new highway and tunnel system. In November 1990, Lazard Frères recommended to Salvucci that the Turnpike Authority be selected to operate the completed portions of the Big Dig because the Turnpike Authority "represents the public entity best suited to operate the Project and the System."

Change at MassPike: Salvucci Takes Control

DUKAKIS HAD LONG WANTED to remove the aging Driscoll from his post, and finally got the Turnpike Chairman to resign in 1987. Dukakis replaced Driscoll with former state senator Allan McKinnon, a man with little apparent interest in transportation policy who undertook the job without any sense that he was being asked to lead the agency that would grow into the manager/owner of the Big Dig. But Dukakis and Salvucci, now for the first time, truly controlled the Turnpike Authority and Salvucci began quietly to restructure it internally to meet his Big Dig plans.

Salvucci soon discovered to his frustration that McKinnon was not keen on his plan for MassPike and the Big Dig. McKinnon's political instincts told him that placing a disproportionate financial burden on the Turnpike Authority would saddle turnpike users with a bill they would pay in tolls for years to come. He questioned whether such a scheme was equitable or politically feasible, but he was a lone voice.

McKinnon also resented the fact that his senior staff was, for the most part, carefully chosen by Salvucci for their talent and loyalty to his broader vision. McKinnon had expected to stay in the State Senate and move up its leadership ladder. When that failed to happen, he actively sought the position of Turnpike

Chairman. Now that he occupied the powerful seat held by the legendary Callahan, McKinnon was loath to be perceived as a simple functionary for Dukakis and Salvucci. In less politically charged times, McKinnon might have had his way. But after Michael Dukakis's defeat for the presidency in 1988, and with his term as governor set to end in January 1991, Salvucci wanted the Turnpike Authority's managers to hold the fort under a possible Republican administration after he and Dukakis left office. In the end, being Turnpike Chairman, with its generous perks and privileges, satisfied McKinnon's ambitions and he never directly interfered with Salvucci's large measure of control over MassPike policies and influence over the selection of the Authority's senior staff. As he neared the end of his term, Salvucci became more comfortable with the possibility that the Turnpike Authority would own and operate a large portion of the completed Big Dig.

Race to the Finish Line: Salvucci Leaves Office

MICHAEL DUKAKIS'S ANNOUNCEMENT in January of 1989 that he would not seek another term as governor was not unexpected, but it changed the political landscape in Massachusetts. Dukakis had been governor longer than any person in the state's history—three four-year terms. For Salvucci, Dukakis's announcement meant that he had a mere two years to get all the remaining pieces of the Project in order. He did not let grass grow under his feet.

Salvucci had won support from the federal government, from Boston's business leaders, and from community activists. He also needed to ensure support among the city's powerful unions, and from the state's Secretary of Environmental Affairs. Salvucci required the Project to develop and sign a Project Labor Agreement, essentially an agreement to hire only from union shops. In return, organized labor promised not to disrupt the Project with strikes.

The most important milestone remaining was the certificate of approval by the state's Environmental Affairs Secretary, which was a precondition to the commencement of construction. Given the scope of the Project, there were many environmental issues that needed to be resolved before the secretary, John DeVillars, would issue his approval. One key element of securing that approval was Salvucci's agreement with the environmental advocacy group, the Conservation Law Foundation (CLF). Just as a strike could cripple the Project, a lawsuit from the CLF could thwart Salvucci's entire elaborate plan. CLF had pushed for a number of significant public transportation improvements to be made in connection with the Project, including substantial new investments in mass transit, commitments to parking freezes, higher tolls and parking fees—tactics to increase the attractiveness of using public transportation. Salvucci was highly sympathetic to incentives to use public transportation, so it was not difficult for him to agree to CLF's terms. He also could not afford the delay that would be caused by a CLF lawsuit, or the mischief that Project opponents might make if they forced the Project to defend its environmental impacts in state court. Time was decidedly in favor of agreement.

DeVillars incorporated the CLF agreement into his certificate approving the Project's Final Environmental Impact Statement and Report, which he issued on his last day in office, January 2, 1991. The certificate included 1,500 separate commitments ranging from noise reduction to rat and dust control to the creation of significant acres of new or reclaimed open space and the commitment to making substantial improvements to the public transportation system.

As 1990 drew to an end, the Project's future remained uncertain. Salvucci had done everything he could to position the Project as an inevitable undertaking. Yet the truth was that not one inch of ground had been moved. FHWA's Record of Decision—the final official act of Project approval—was still five months away. A new administration under Republican Governor William Weld would soon take office, and with the economy in a recession, it was not beyond the realm of the possible that Weld would take a hard second look at the Project as envisioned by Salvucci. Weld's selection of James Kerasiotes to be the new Highway Commissioner—the same Kerasiotes who had been a loyal aide to Governor King's Transportation Secretary—raised the specter of Ed King and the old third harbor tunnel alignment. Would Kerasiotes, a King man to the core, scuttle the Artery depression and simply build a new tunnel and the Leverett Circle connector?

Salvucci left office hoping that the coalition he had built would hold, hoping also that his continued power over MassPike and Massport would afford him some leverage. But the future of the Project was very much in doubt when Michael Dukakis walked down the State House steps early in the afternoon of January 2, 1991.

CHAPTER FIVE

Building the Big Dig: The Kerasiotes Years

FRED SALVUCCI THOUGHT, ever so briefly, that Governor Weld might ask him to stay on as Transportation Secretary so he could guide the Project through construction. Dukakis had given Salvucci an eleventh-hour appointment to the Massport board, but that only gave him limited ability to influence transportation policy. He spoke with both Weld and Lieutenant Governor Paul Cellucci, advising the incoming leaders to designate their choice for owner of the Artery/Tunnel Project as soon as possible. "You really want the entity that will be having to operate and maintain the facility at the table during design and construction decisions," recalled Salvucci.

Salvucci recommended that the responsibilities be allocated between Massport and MassPike, but his advice was not taken. The Weld Administration did not engage the question whether Massport or MassPike —or a combination of the two—would become owner/operator of the completed Big Dig until the Dukakis loyalists managing the Turnpike Authority forced the issue in 1995.

Weld replaced Salvucci with Richard Taylor, the state's first African American Transportation Secretary, but astute observers understood that the real power behind the Big Dig was not Taylor but the new Highway Commissioner, a man with no relationship to Taylor, who had served in the King Administration and had now returned to reclaim his power and his hold on state transportation policy.

Jim Kerasiotes was a man driven by power and ambition. His ultimate goal was to be named chairman of the Turnpike Authority, where he could become the master builder of his generation. Ironically, it was the great platform Salvucci had built that enabled Kerasiotes to fulfill his life's great ambition. Kerasiotes could easily have persuaded Weld, faced with an economic crisis and personally unengaged by transportation issues, to scuttle the Artery and return to simply building the third harbor tunnel. But Kerasiotes believed in the Project, and in his destiny to be its builder, and he was determined to go forward with it.

Kerasiotes was personally devoted to Ed King and to the former governor's "can do" philosophy. As expressed through Kerasiotes' temperament, this approach to governing appeared highly partisan, but in fact he never looked at the political world as a Republican loyalist. Kerasiotes was a smart and pragmatic official who rarely lost sight of his political and policy objectives. He made his most important deals with Democrats and crafted the final legislative agreement to formally vest Big Dig responsibilities in the Turnpike Authority with McKinnon's General Counsel, a Salvucci loyalist.

Kerasiotes enjoyed his self-crafted reputation as a tough, unsympathetic, brash, plain-talking leader. In an opinion piece for the *Wall Street Journal* Kerasiotes articulated his approach to managing in the public sector: "Get your team on the field. If you have hold-over managers who aren't on the same page, move them out." He remembered how Dukakis-era holdovers had stymied King's agenda; he was determined not to let them do it again. In the last analysis, he believed that fear was a potent motivating force—and he would rather be feared than liked.

Kerasiotes encouraged his tough-guy reputation by making outrageous proposals and comments: he wanted employees in the MBTA fare-counting room to have their pants pockets stitched; he told an apocryphal story of a Highway Department employee who threw a stuffed animal out a car window and then requested overtime pay for having to remove a "dead animal" from the roadway. He was said to display a hatchet prominently in his office as a not-too-subtle message to those who were not on his team, and there were accounts of him walking through corridors firing people he did not trust or have confidence in. These stories got Kerasiotes a disproportionate share of publicity and solidified the image he believed would instill fear in some, respect in others. This outward expression of bravado was, for him, largely calculated and part of "the game," but it left deep wounds in many people who, when he needed support later in his career, were not inclined to offer him a helping hand.

Kerasiotes' disdain for Dukakis and Salvucci was palpable, but his feuds with peers in his own administration—notably Massport Director Steven Tocco and Weld's chief of staff Virginia Buckingham—revealed that his disdain for others whose abilities he questioned, or whose policies he objected to, crossed party lines. He was a singular personality, a complicated man who listened to his own drummer and did not suffer fools gladly. When *Boston Globe* columnist Brian Mooney referred to him as "arguably the most hated man in state government," one of Kerasiotes' admirers called the columnist to ask: "Who's arguing?"

Kerasiotes and MassPike: Round One

WHEN KERASIOTES BECAME Highway Commissioner in 1991, he quickly locked horns with the Turnpike Authority, still run by Dukakis appointees. MassPike Chairman Allan McKinnon's term would not expire until July 1996, and he and his senior staff were determined to make a significant investment in rebuilding the turnpike roadway and particularly the deteriorating Sumner and Callahan Tunnels. Ceiling tiles in the Callahan Tunnel periodically fell onto vehicles passing below, and the Sumner Tunnel's ceiling support system was literally falling apart, having not been reconstructed since it was built in the 1930s. No major reconstruction had taken place on the 135-mile turnpike since it was built, and routine maintenance activities could no longer keep up with the wear and tear caused by heavy traffic and harsh New England weather.

Turnpike officials also undertook the unpopular task of raising tunnel tolls (1989) and turnpike tolls (1990). These toll increases were the first to be implemented in over a decade, and they would enable the Authority to pay for its rehabilitation program without disabling its ability to provide necessary funding for the Artery/Tunnel Project when that was necessary. Against Weld's and Kerasiotes' wishes, the Turnpike Authority embarked on a $500 million rebuilding effort. Weld, faced with a serious economic crisis, wanted access to MassPike's revenues to help balance the state budget.

A bitter political battle took place, but Turnpike officials had done their political homework, and the reconstruction program went forward. Kerasiotes told Thomas Kiley, a former First Assistant Attorney General and friend of McKinnon's General Counsel, that he "wanted the message to go back that he [Kerasiotes] was going to turn up the heat every day until either

McKinnon was out of business or the world blew up." At one point, Weld and Senate President William Bulger hatched an elaborate scheme to appoint McKinnon to a state college presidency and the Turnpike's General Counsel to a judgeship—removing two perceived obstacles to the Weld/Kerasiotes agenda—in return for Weld's removal of State Treasurer Joseph Malone, a Bulger opponent, from authority over the Convention Center, a Bulger sinecure. The attempted deal blew up when it was leaked to the *Boston Globe*, and McKinnon was satisfied to see out his term, particularly after his General Counsel quietly secured him a significant pay raise through the legislature.

Once it became clear that McKinnon would finish his term, it was inevitable that the common interest in the Big Dig shared by Kerasiotes and Salvucci's MassPike loyalists would bring the feuding parties together. By 1994, Kerasiotes began to reach a rapprochement of sorts with the Turnpike Authority. Discussions between Pat Moynihan, Kerasiotes' loyal deputy, and the Turnpike's General Counsel led to the development of crucial pieces of legislation guiding and directing the transfer of the new tunnel from the state

Keeping the harbor at bay: the cofferdam that allowed the third harbor tunnel to be built. Logan Airport can be seen across the harbor.

Highway Department to MassPike. This legislation, passed in the summer of 1995, marked the first time that the legislature officially authorized the Turnpike Authority to be the owner and operator of the Project. And it marked the end of the long effort to resolve the ownership issue, serving the needs of both Kerasiotes, who anticipated running the Turnpike Authority when McKinnon's term expired in 1996, and the Salvucci holdovers at the MassPike, who wanted to ensure that the Lazard Frères plan was put into place before they left office.

The 1995 legislation also required MassPike to make a $100 million contribution to the Project—the first of many financial contributions MassPike would make in an effort to ease the overall burden on the state's Highway Fund and enable Massachusetts to continue to spend about $400 million each year on construction projects throughout the state. By 1997, MassPike's total contributions to the Project would exceed $1 billion, and ultimately would approach the $2 billion mark. With the Turnpike Authority making these substantial contributions to the Project, not once in the years following 1990 did the state have to raise its gasoline tax to pay for the Big Dig or other state highway projects—a fact that the Project's critics rarely noted.

Building the Big Dig: The Third Harbor Tunnel

WITH ALL THE POLITICAL ELEMENTS aligned by Salvucci, and with Kerasiotes prepared to be the Project's hard-driving builder, construction of the third harbor tunnel began. The tunnel would be the first element of the Project to be completed. Building a tunnel under an active harbor presents daunting engineering, construction, and environmental challenges. In this instance, the Project chose a construction method never before used in Boston, the connection of prefabricated immersed tube tunnel sections into one seamless whole. The process began with the removal of 900,000 cubic yards of rock, clay, and polluted silt. The silt and sludge were used to fill what was left of old Governor's Island to complete an expansion of the airfield at Logan Airport. The remainder was shipped off to the deep ocean and released at an approved disposal site. A three-quarter-mile-long trench one hundred feet wide and fifty feet deep was then dug into the harbor floor. Huge sections of bedrock needed to be blasted away before the trench could be dug. In one of the extraordinary steps taken to construct the Project in an environmentally

From Baltimore to Boston: the immersed tunnel tubes that would become part of the new harbor tunnel.

sensitive manner, $1 million was spent for a "fish startle" system meant to scare fish out of harm's way before blasting.

At the same time, twelve 325-foot steel immersed tube tunnel sections were built at the Bethlehem Steel shipyard near Baltimore and floated up the coast to South Boston. There they were fitted out in assembly-line fashion and, when finished, filled with enough concrete and steel to allow each section to sink into a pre-dug trench. A significant amount of work then took place underwater. Divers used laser lights and global positioning units to guide each tunnel section into place, then—by means of a complicated process involving couplers and hydraulic jacks—tunnel sections were connected with an airtight seal. The first tube was lowered in February 1993, and by the end of 1994, all twelve sections were in place.

The connecting roadways on the South Boston and airport sides of the new tunnel posed their own construction challenges. Because much of Boston's land mass is man-made, and therefore prone to instability, constructing the tunnel approaches required advanced deep-soil mixing techniques. Buried in bedrock eighty-five feet deep on the South Boston side was the largest cofferdam in North America, 250 feet in circumference with thirteen foot walls.

Inside the third harbor tunnel

Kerasiotes was determined to open the tunnel as early as possible. Governor Weld was planning to run for the United States Senate against incumbent Senator John Kerry in 1996. Opening the new tunnel just before election year would be a boost to Weld and another feather in Kerasiotes' cap. In the effort to open the tunnel on a fast schedule, several decisions were made that would later haunt the Project. Among them was the installation of a temporary system of smoke, fire, and incident detection units operated by a suite of software that was built in two phases.

When the developer of the software for phase one failed to win the bid for the phase two contract, it claimed ownership to the source code—the essence of the software—and refused to allow the phase two contractor access to this essential information. That dispute was resolved following extensive negotiations and threatened litigation. Then the phase two contractor demanded more money for alleged costs associated with delayed access to the source code. While it can certainly be said that these private-sector companies did their best to take advantage of the Project, it was also true that by phasing the software development contracts—something required by the early-opening approach taken by Project managers—the Project was making itself particularly vulnerable to such unscrupulous contractors.

On the day the new tunnel was opened, its ownership was transferred from the state's Highway Department to the Turnpike Authority. This transfer from a Kerasiotes-controlled agency to a Salvucci-controlled agency was a significant political achievement, for it marked the first public expression of true bipartisan cooperation in the operation and management of the Project. Salvucci's plan was still in place, and Kerasiotes, who expected to be leading the Turnpike Authority as its new chairman in July 1996, was more than happy to begin the process of integrating MassPike into the Artery/Tunnel project.

Building the Big Dig: Working below the Central Artery

WHEN FRED SALVUCCI first considered rebuilding the Central Artery underground, his thoughts turned quickly to the daunting task of building a new highway without disturbing the existing one. Keeping Boston's traffic moving during construction would be a nonnegotiable requirement. Initially, Salvucci thought the challenge insurmountable. In time, advances in construction technology made the unthinkable possible. By building a series of slurry walls alongside the existing elevated Artery structure, new support beams could be put into place, the old support system dismantled, and the tunnel below the Artery dug and built out with minimal disruption to the surface streets above. The Central Artery could continue to function without interruption during the construction of its replacement tunnel directly underneath it.

The use of slurry-wall construction was the Project's most important innovative technique because it enabled city life to continue largely uninterrupted while the massive digging project went forward: the Project could dig under the existing Artery without interfering with the flow of vehicular and pedestrian traffic above. The slurry walls eventually would become the actual exterior walls of the new underground Artery. Project builders first dug a trench, removed the earth, and pumped a liquid slurry mixture (a combination of clay and water) into the excavated hole. Steel reinforcing rods were placed into the slurry, and then the slurry was pumped out as concrete was pumped in to replace it. The final result: a solid underground concrete wall along the entire length of the new interstate highway.

But before any underground construction could take place, the Project needed to overcome yet another challenge posed by the old city: relocating a complex, poorly documented maze of underground utilities. Over the years,

Building the new highway piece by piece: a new viaduct leading to the underground artery

Boston's underground utilities had been put into place in a haphazard manner. The untidy collection of pipes, lines, and conduits carrying electricity, water, gas, steam, and sewage was so old that the utility companies themselves did not know their precise locations. Construction of the new underground Artery required the relocation of over twenty-nine miles of utility lines. In addition, 5,000 miles of fiber-optic cable and 200,000 miles of copper telephone cable were installed, providing the city and region with one of the most important and little-known benefits of the Project: a renewed, modern public utility network through the old downtown core.

There were other challenges. The underground Artery would have to be built along an alignment that included two old transit tunnels: the Blue Line tunnel and the Red Line tunnel. This required careful design and engineering, especially at the Dewey Square intersection in front of South Station. This was the lowest point of the underground Artery—120 feet beneath Dewey Square. It was also the most crowded point, a compact area that would have to accommodate both new highway and transit components.

Salvucci and his planners designed a magnificently complex underground transit and highway "city"—a layer cake whose base was a four-lane interstate highway, passing beneath the almost century-old Red Line tunnel, on which

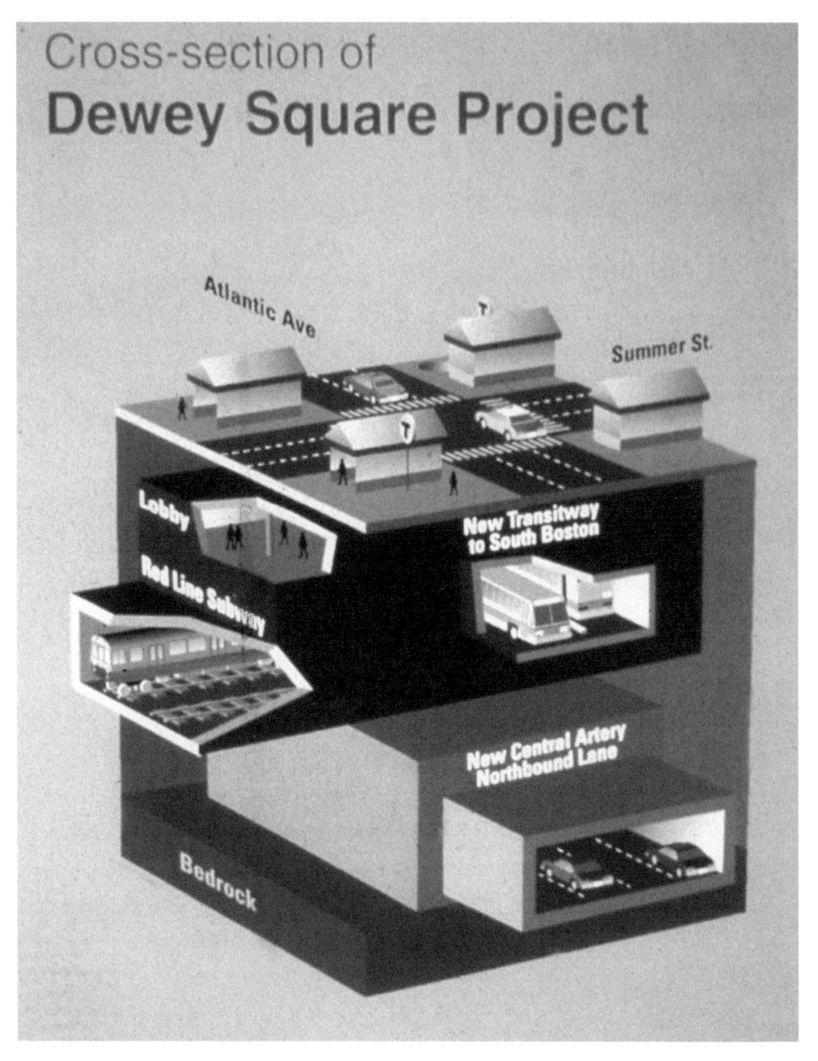

Boston underground: the complexity of the Big Dig at Dewey Square

the new Silver Line transitway rested, and all topped by a modern lobby for the transit users. Salvucci was committed, beyond all else, to improving the regional public transportation system, and he looked for ways to do so in the context of the massive Central Artery/Tunnel highway rebuilding project. One hundred and twenty feet below the surface of the street, workers built a "bridge" under the Red Line by constructing an elaborate system of tunnels, shafts, and grouting gallery tunnels. This so-called underpinning process enabled the existing Red Line tunnel to have sufficient support to avoid dangerous settling as a result of the building of the new highway tunnel above.

Another major challenge was the construction of a complicated series of tunnels and ramps in the area behind South Station—an area filled with active Amtrak and commuter rail lines. The Project could not disrupt the operation of these important transportation systems. The solution became the largest tunnel-jacking project in the world. In order to place the new highway tunnels under the existing active rail lines, the Project literally pushed prefabricated tunnel boxes through the earth around the rail lines by the use of hydraulic jacks. More than fifty jacks were used, each with a maximum pushing capacity of 10,000 pounds per square inch. In order to ensure the stability of the earth, the ground around the tunnel-jacking area was frozen. A large number of pipes were driven into the ground and filled with a brine solution that was passed through refrigeration units in order to keep the ground frozen, turning the earth around it as hard as rock and allowing excavation and jacking to occur without the dangers of ground settling and collapsing.

Digging in the downtown core of a city more than 370 years old led to the discovery of a treasure trove of archeological finds. While history was being made, history was also being rediscovered, and the Project made a special effort to preserve and catalogue thousands of artifacts that will provide historians for years to come with a new window on the city's past. Colonial-era artifacts ranging from pottery and glassware to musket balls and children's toys were released from their long captivity under the streets of Boston. In Charlestown, construction workers uncovered the site of John Winthrop's original encampment. And on Spectacle Island, Project builders discovered a Native American campsite that led to important insights into the ancestry of the Neponset tribe.

Building the Big Dig: The Fort Point Channel

BUILDING AN ELEVEN-LANE interstate highway under the narrow Fort Point Channel proved to be the most daunting part of the entire Project. The 1,100-foot crossing had such massive and unexpected construction problems that its cost skyrocketed to $1.5 billion—easily the most expensive section of interstate highway ever built in America. The highway had to cross under the Fort

The Fort Point Channel dry dock, big enough to build an aircraft carrier

An aerial view of the Fort Point dry dock, with the seaport access road visible in the distance

Point Channel at this particular place in order to accommodate the alignment of the third harbor tunnel, which had to open out directly onto Logan Airport and not in or through East Boston.

Salvucci's idea for the Fort Point alignment had come after another conversation with Bill Reynolds, who was working on an unrelated project near South Station. Reynolds called Salvucci and the two men looked at the area of South Boston across from the Fort Point Channel that inspired Reynolds's insight. Knowing both the political and engineering obstacles to the tunnel portion of the Project, Reynolds told Salvucci, "If you go through East Boston, Tip O'Neill is going to stop it." On the other hand, if Salvucci built the tunnel from downtown and tried to "swing over to the airport, the curve is so violent that Federal Highway is not going to like it. . . . The right way to do it is to come across South Boston and over to the airport." Salvucci recalled that "you can physically see [the alignment] because that section of South Boston is basically underutilized parking lots and rail yards. It is a vacant swath right across South Boston." As he observed the area and took it all in, he could see it as clearly as Reynolds, as if the plan were literally unfolding before his eyes. Salvucci turned to Reynolds, and in his typically understated way, said, "Gee, you are right."

Boston's Fort Point Channel is a shallow, muddy inlet, virtually all the water that remains of Boston's South Bay. This remnant of the days when Boston was isolated on the Shawmut Peninsula now is surrounded by man-made land. The Channel serves adjacent water-dependent and industrial uses, foremost among them the world headquarters of the Gillette Company. Gillette's Fort Point plant manufactures over one billion razor blades every year—blades that require the use of precision instruments. Construction under the Channel that disturbed Gillette's manufacturing could have a devastating impact on one of the nation's leading corporations. Moreover, the Red Line transit tunnels passed under the Channel almost exactly where the proposed highway alignment would go.

In addition to these logistical and environmental challenges, Project engineers encountered problems as soon as construction began. The first was a whopper: when digging began, they discovered that the Channel's soil conditions were significantly more unstable than they had predicted. Engineers had assumed that they could use two immersed tube tunnels (similar to the tunnels used in the construction of the third harbor tunnel) in the center of the Channel, connecting on either side to less expensive "cut and cover" tunnels.

The unstable soil made "cut and cover" tunnels impossible, and the engineers had to revise their plans to include six immersed tube tunnels. Some engineers traveled to Japan to receive training in deep-soil mixing techniques developed and used successfully in unstable soil conditions in that country.

But the need to sink six immersed tube tunnels in the Channel posed another challenge. The Channel was too narrow for these tunnel sections to be fabricated off-site and floated into place. The tunnels would have to be built on-site. To do this the Project constructed an immense dry dock in the Channel. At 1,000 feet long, 300 feet wide, and 60 feet deep, this dry dock was enormous enough to accommodate construction of the nation's largest aircraft carriers. Mindful as well of the need to preserve worthwhile elements of the old city's historic infrastructure, before building the dry dock the Project carefully dismantled a hundred-year-old seawall piece by piece, and rebuilt it when the dry dock was no longer needed.

A series of six cofferdams kept the Channel water out of the dry dock and enabled construction of the tunnels sections. When sections were complete, they were floated out into the Channel and sunk into place. Project construction workers, some of them underwater, ensured that the submerged tunnels did not interfere with the Red Line—no small matter since at certain points the bottoms of some new tunnel sections were just six feet above the Red Line ceiling.

In the end, the Fort Point Channel crossing was not only an important but indeed an essential component of the overall Project. But its inflated cost contributed significantly to the pressures Kerasiotes faced as he attempted to keep the lid on the Big Dig's budget.

Building the Big Dig: The Charles River Crossing

LIKE THE FORT POINT CHANNEL crossing, the Charles River crossing was enormously complicated, but it further required not only sophisticated engineering but also an arduous political process to satisfy public concerns about its aesthetic impact on the city. The river crossing marked the point at which Interstate 93 connects with the Tobin Bridge, Storrow Drive, Leverett Circle, and points north. Project engineers developed thirty-one different approaches—or schemes—for crossing the river, schemes that contemplated tunnels, bridges, or combinations of the two. Salvucci chose an "all bridge" solution named Scheme Z that proved enormously unpopular. Scheme Z's

Christian Menn's spectacular bridge rises above the Charles River.

The Zakim Bridge and Leverett Circle connectors. Scheme Z is seen in the background.

high ramps and complex design, including an eighteen-lane bridge resting on seventeen piers, quickly became the target of Cambridge and Charlestown neighborhood activists. They believed that Salvucci had not given the river crossing the same sensitive, careful review he had accorded other elements of the Project.

Salvucci stubbornly refused to give way, even in the face of opposition from some of his own designers. In his view, Scheme Z may have been big and ugly, but so was the area it would run through. The crossing—at the point where the Charles River empties into Boston Harbor—was proposed for an area that historically had been an industrial wasteland. Nevertheless, many saw Scheme Z as an unacceptable solution to the problem—an ugly path of least resistance. Scheme Z came under a pounding by environmental activists and a collection of self-styled design experts and Project opponents. When *Boston Globe* transportation reporter Peter Howe made Scheme Z the target of several days of negative stories, the writing was on the wall that the design approved by Salvucci would not pass muster. Lawsuits were threatened, and the *Globe*'s architecture critic labeled the Scheme Z a "Great Wall" of concrete over the Charles.

When William Weld became governor, he pledged a review and reconsideration of Scheme Z. Transportation Secretary Richard Taylor put together a large and unwieldy commission to study various alternatives to the river crossing. One critic came up with an all-tunnel design alternative that would cost the Project an additional billion dollars. The commission struggled, as Salvucci had, to find a way to cross the river through one or more tunnels and not break the Project budget. Taylor's bridge committee, a subgroup of the larger commission, finally recommended a new approach to crossing the river with the completely forgettable name CIP 8.1D, Modification 5 ("Mod 5"). This recommendation included a three-lane tunnel as part of its solution, but also a new bridge design—an elegant structure designed under the supervision of the Swiss bridge designer Christian Menn. Within weeks of the Mod 5 recommendation, Taylor was gone and James Kerasiotes was the new Transportation Secretary.

True to form, Kerasiotes would not consider a solution that substantially increased Project costs simply for aesthetic reasons. Instead, he made marginal changes to the original "bridge only" Scheme Z design, including more parkland and open space for the neighborhoods, and he embraced the Menn bridge design—a design so stunning that it effectively silenced most of the loudest critics. The brilliant, beautiful bridge design saved Scheme Z and also saved substantial Project costs. In the end, despite years of struggle and opposition, the Project opted for building a solution almost identical to the original Scheme Z with one important difference: Menn's outstanding cable-stayed suspension bridge. Kerasiotes pushed forward a river-crossing approach that would not bust the budget, would provide appropriate mitigation to surrounding neighborhoods, and would provide the city with the one piece of transportation architecture that would become emblematic of Boston. "The federal government is our partner and our banker," said Kerasiotes. "We need to convince them of our commitment for integrity."

It was an example of Kerasiotes' approach to problem solving, which could be very effective. He was not interested in sitting for hours listening to experts and advocates explain their proposed solutions to Project problems. His style was to bring the matter in house, close the door, identify the options, and come up with a politically palatable solution. It was an approach that only the tough and confident Kerasiotes could pull off, and in this instance it worked and it enabled the Project to move forward.

CHAPTER SIX

The Struggle to Complete the Project: From Kerasiotes to Amorello

ALLAN MCKINNON'S term as Turnpike Chairman expired on June 30, 1996. It was no secret that Governor Weld's choice for his successor would be James Kerasiotes. It was the fulfillment of Kerasiotes' ambitions, and for the next year he served both as Transportation Secretary and Turnpike Chairman. He held on to both positions until he transferred all of MassHighway's power over the Project to the Turnpike Authority. Once he believed that he could control the Project as chairman of the Turnpike Authority, Kerasiotes relinquished his position as Transportation Secretary. He was master of the state's transportation world, and he loved every minute of it.

His inner circle was loyal and hard working. Kerasiotes relied in particular on Patrick Moynihan to be the "good cop" in dealings with other state officials. Moynihan would move about the transportation secretariat, as deputy to Kerasiotes, as general manager of the MBTA, as Secretary of Transportation, and finally as project manager for the Big Dig.

Kerasiotes moved quickly to consolidate his power over the Project. In July 1996, he quietly signed a "Project Management" agreement with the Highway Department, which he still controlled. Through this agreement, the Highway Department essentially transferred all its powers and authority over management of Artery/Tunnel construction over to MassPike. This was extraordinary, particularly since no state law at that time gave MassPike power over the construction of the Project. Legislative leaders grumbled, but did nothing. The truth was that they were more than happy to have Kerasiotes take responsibility for the Project, as they feared both its complexity and potential costs. In January 1997, the legislature passed into law a recodification of the Turnpike Authority's enabling act, creating a "metropolitan highway system" and vesting the Turnpike Authority with nearly complete power over the new system.

With his power consolidated and his authority unquestioned, Kerasiotes accepted no criticism and conceded little ground to those who, in his words, "view the Project as an endless opportunity." At one of his annual "State of the Project" briefings he told a group of civic leaders, "We have always tried to be realistic about what's possible and what isn't. This approach has gotten us into a few wrestling matches. But it has kept the Project in balance. That is our job, and we don't apologize for it."

Kerasiotes' "wrestling matches" included his refusal to agree to an eminent-domain taking of the Spaulding Rehabilitation Hospital—a $90 million savings for the Project that was tested in court, and won by Project lawyers—and his refusal to give way to the environmental activists who continued to demand an expensive "tunnel only" solution to crossing the Charles River. No one could accuse Kerasiotes of being soft on Project costs, but as each year went by, it became increasingly clear to some Project officials that they could not continue to hold the line.

The Bill Comes Due: The End of the Kerasiotes Era

KERASIOTES MANAGED THE BIG DIG during its early and peak construction years—when 5,000 construction workers were at work on the "Big Dig," and the Project was spending up to $3 million every day. During that time, traffic moved smoothly through the city, and the Project maintained an extraordinary construction safety record.

He faced his first Big Dig financial crisis in 1994, when Bechtel's project manager publicly undermined Kerasiotes' "on time, on budget" mantra. The Big Dig's budget numbers "are very, very unpredictable numbers," said Bechtel's Tad Weigle, Jr. "You just don't know." Weigle had privately informed Project officials of his belief that the Project would end up costing about $13 billion. This was far more than the $10.4 billion number that represented the official Project budget that year, and certainly much more than Kerasiotes at the time believed would be the actual cost of the Project. The Bechtel manager was asked to leave the Project, a clear signal that Kerasiotes would not accept public dissension from his consultants or disloyalty to his approach to keeping the lid on Project costs.

Kerasiotes had a simple philosophy about Project costs: they would increase if you created a fertile environment for them to increase. Public predictions of higher costs would become self-fulfilling prophecies, because the construction industry would simply take those predictions as the new "floor" of Project costs and make them a reality. Salvucci agreed with him in this philosophy, telling the *Boston Globe* that there were "hundreds and hundreds of people who have to make decisions in time to get something to happen, and somewhere along the way there's always going to be some slippage. If you don't hold, if you don't portray what you think is realistic but tough . . . then you're sending the signal, 'hey, cost is no object, time is no object, mañana is OK', and you'll totally lose control of this thing." Kerasiotes was determined to set what he believed was a credible cost, and hold to it.

When Salvucci left office in 1991, the Project's estimated cost was nearly $6 billion. As a result of the changes in Scheme Z, the costs of inflation, and the difficulties encountered in the Fort Point Channel crossing, by 1996 total project costs had been pegged at $10.8 billion—a "net" number that included

Patrick Moynihan and Jim Kerasiotes

within it several assumptions, including a significant recovery of cash from the Project's innovative Owner's Controlled Insurance Program. Kerasiotes made the budget sacrosanct: the Project, come hell or high water, would come in at $10.8 billion, and everyone had to manage to that budget figure. If costs increased in one place, they had to be brought down in another. In March 1999, he told the *Boston Globe*, "Everyone knows at the Project there is zero tolerance for failure to achieve a milestone." Reports of pressure on the Project's budget and schedule did not trouble him. "Through seven-plus years, we've delivered when we said we were going to deliver," said the ever-confident Kerasiotes.

In this environment, Bechtel's manager in charge of Project finances, Bill Edwards, invented what became known as an "up/down" chart, a periodic demonstration of those areas where Project costs were rising, and those other areas where Project managers were prepared to make cutbacks to offset the increases. On a regular basis, Project officials pored over Edwards's new "up/down" chart, identifying causes of cost increases and ways to reduce costs in order to manage in a "zero sum game" environment. Ominously, Edwards began to create additional charts that he labeled "Armageddon" charts.

Bechtel and its joint venture partner Parsons Brinckerhoff established a "Board of Control" to help manage the Project. The Board of Control was a small, almost secretive group, made up of high-level company and Big Dig officials who met regularly, dined well when they met, and monitored all aspects of the Project. At a Board of Control meeting on July 22, 1999, that included in attendance the president of Bechtel Infrastructure and the CEO of Parsons, Kerasiotes spoke of his concern that "the dedication of all parties to the budget may have waned a bit." Despite these fears and warnings, Project managers continued to believe that their monthly "up/down" exercise was sufficient to manage the increasing pressure on the budget.

Kerasiotes took a number of steps, from the purely symbolic to the significant, to keep Project costs down. He scaled down or removed the aesthetic finishes for the underground tunnels and ventilation buildings, and he eliminated a ramp that would service traffic from the new tunnel to Chinatown and the Back Bay. He adamantly refused even to consider the city's proposal that he build a "slingshot" ramp to get traffic into the Back Bay. He said provocatively that the Back Bay already had a ramp. "It's called Storrow Drive."

Kerasiotes was at the peak of his power when he spoke before a packed ballroom at the Park Plaza Hotel on October 7, 1999. That day he was opening the new Leverett Circle connector—the one Project idea that Ed King had previously embraced—and his pride was palpable. His years of work were beginning to bear fruit. His management focus was on keeping the city moving and fully functional during the Project's peak construction years. This he did, with great success. But the Project was an organic, dynamic thing, a multifaceted behemoth that required daily focus and attention. As Kerasiotes became more and more comfortable with his position, his attention to detail suffered, and he let his mastery over Project finances slip.

Patrick Moynihan took on the demanding job of Big Dig Project Director in January 1999. Moynihan did not realize that he was inheriting a long-brewing financial crisis—a crisis hidden from Kerasiotes by functionaries too intimidated to bring bad news to him. A financial time bomb was ticking, and Moynihan was sitting on it. He reacted initially with anger toward his friend and predecessor, Peter Zuk. Moynihan privately took Zuk to task for not working with Kerasiotes "to find additional revenue sources or introduce reasonable options to reduce the scope of the Project." Moynihan declared his determination to "work to contain Project costs" even though "we have been left a tool bag that is virtually empty." Zuk, for his part, took issue with Moynihan's assertions, reminding him of Kerasiotes' "disinclination to know about the cost of the Project." In a draft memorandum to Kerasiotes prepared for Moynihan by Bechtel on the occasion of his first six months on the job, Moynihan ruefully noted that the Project had "an assertive definition of 'aggressive but achievable,'" and that he would plug ahead because his "short stint on the Project has taught me to embrace the impossible, and aggressive results will be achieved."

Mischief was in the air. State Treasurer Shannon O'Brien had previously locked horns with Kerasiotes over Artery funding. In late 1999, O'Brien began to demand more support for Big Dig financial data that was included in state bond offerings. Moynihan was asked to sign off on state bond prospectuses that included Big Dig cost estimates that were increasingly out of touch with reality. Moynihan was put in a difficult and untenable position: he understood that the $10.8 billion figure was probably no longer viable, that the harsh reality was that the increasing pressures on the budget could not be solved with more cuts, or simply with Turnpike Authority revenue, but he had not been able to explain the situation to Kerasiotes effectively. In late November 1999, Moynihan initiated a process that would include a "bottom-up" review of all Project costs and gradually bring Kerasiotes to an understanding of the full dimensions of the crisis. His effort began too late.

As Moynihan was preparing to brief Kerasiotes, the Turnpike Chairman was engaged in a journalistic frolic that would cost him dearly. The *Wall Street Journal*, trying to boost local sales, had established a New England edition and hired aggressive reporters to scout out fresh stories that the *Boston Globe* and *Boston Herald* had missed. In a series of unguarded interviews with the *Journal*'s Geeta Anand, Kerasiotes let his hubris get the best of him. He said things about himself, Governor Paul Cellucci, and others that would find their way into print at just the wrong time. Anand was working on a number of Big Dig stories, and she kept the Kerasiotes interview handy for an opportune moment.

Moynihan and a small group of trusted advisors briefed Kerasiotes on December 15, 1999, in a conference room at the new Seaport Hotel. Kerasiotes pushed back on his advisors and insisted that the numbers be scrubbed hard before he would go public with a cost-increase announcement. He also wanted to begin the process of getting his political ducks in line, but he seriously miscalculated the speed at which events were moving. Another state bond issue was imminent, and that meant another round of disclosures to the state treasurer and others. Kerasiotes made his announcement of an additional $1.4 billion in Big Dig costs on February 1, 2000—the same day the Federal Highway Administration had conditionally accepted a finance plan that was still pegged at the now defunct $10.8 billion. The impact on federal and state officials was breathtaking. "How did it happen?" asked *Boston Globe* transportation reporter Thomas Palmer. "Nothing had indicated anything of this nature," said the vice chairman of the Wall Street credit rating agency Fitch IBCA.

Adding substantial fuel to the fire, Geeta Anand's interview with Kerasiotes ran in the *Journal* exactly one week later. It was a bombshell. The piece quoted Kerasiotes describing the governor's chief political advisor as a "moron" and his former chief of staff as a "reptile." Kerasiotes also declared that the governor himself was afraid of him. David Luberoff, author of a Harvard University study of the Big Dig, reacted harshly to Kerasiotes' remarks, telling Anand, "Here you have a guy who misled everyone, and now it turns out the project is way over budget. You have to ask the question: Why isn't he being fired?"

In Washington talking with FHWA officials and trying to keep the dike from opening wide, Moynihan exclaimed: "Are they all trying to destroy him?" But Kerasiotes was destroying himself, albeit unwittingly. His intemperate comments came at precisely the wrong time, a time when he was vulnerable to attack for having "hidden" the cost overruns. Kerasiotes' pride would never allow him to confess ignorance—although such a confession would have been equally damning to the widely held view that he had known the extent of the cost overrun and hidden it. As Robert Havern, chair of the State Senate's Transportation Committee, remarked, "Was Kerasiotes hiding it, or did he not know? Either answer is not very good."

In the end, Kerasiotes had alienated so many people in so many places that his friends and allies were too few to be of any significant assistance to him when he most needed the help. Under investigation by the Securities and Exchange Commission, Kerasiotes found that he had, over the past several years, offended the Federal Highway Administrator, insulted the governor, and generally alienated the state congressional delegation, the legislature, and the press. Governor Paul Cellucci believed that Kerasiotes' personality, coupled with the looming Big Dig budget overrun, might reverberate against his administration. Congressman James McGovern summarized Kerasiotes' problem succinctly: "The personalities are part of the problem—the reluctance to be open about facts. They have this bunker mentality that manifests itself in outright hostility. When public officials say it's on time and on budget, then it's not unreasonable to expect it's on time and on budget."

Senator Havern quipped that Kerasiotes had "just dropped a grenade with the pin pulled right in the governor's lap." Knowing a loaded grenade when he saw one, Cellucci knew that he had no choice—and he asked Kerasiotes for his resignation on April 11, 2000. Kerasiotes briefly resisted, but he understood that without the confidence of the governor he could not manage in the manner he was accustomed.

His letter of resignation was dignified and brief, asserting that "when the final reckoning is made, my record will stand as a solid one in service to the people of the Commonwealth." Kerasiotes was a bit more combative in a statement he released the day he left office: "Clearly people feel I did mislead them, and I must accept that verdict and move on. But I tell you, in my heart, I feel I pushed back on the budget's bottom line for only one reason: it was how I did business, to push back, to demand." And with that, the master of the Big Dig was gone, replaced briefly by the governor's secretary of administration and finance and then, for two long years, by a renegade Turnpike Authority board of directors led by a part-time chairman.

Kerasiotes may have had the distinction of making the fewest friends and allies of any public official of his times, and it easy to remember him for the crisis of cost overruns that forced him out of office. But it would do both the man and history an injustice to forget that Kerasiotes was, in many respects, the right man at the right time—the Project needed a tough, committed leader in the years following Fred Salvucci, and while the two men would never admit it, they were alike in their single-minded focus on seeing the Big Dig through, despite all obstacles. The Big Dig could never have been conceived and designed and approved without Fred Salvucci; it could never have been built without Jim Kerasiotes.

The Completion of the Project: The Amorello Era

MATTHEW AMORELLO came to the Turnpike Authority in a time of crisis. The Turnpike Authority board, let loose in the aftermath of the Kerasiotes reign, was behaving erratically and, without a steadying hand, appeared set on a course that could jeopardize the completion of the Project. Kerasiotes' successor, Andrew Natsios, undertook a reorganization of the Authority's management structure that was both ill conceived and in violation of the Authority's enabling act. The consequence was that, when Natsios left the Big Dig for a political appointment in Washington, he left behind a huge administrative mess that threw MassPike and the Big Dig into a year and a half of turmoil.

Board members fought publicly with one another and with Bechtel officials, and an elaborate game of finger pointing took up much of the time of MassPike board members and Project officials. Too much was at stake for the governor to continue to allow MassPike to sink under the weight of this discord. The state had put its confidence in the Turnpike Authority as the agency that had the financial capacity and internal stability to guide the Big Dig through completion, and then manage the new transportation system.

Acting Governor Jane Swift understood the importance of bringing a steady hand to MassPike. She attempted to fire two board members, but her efforts were overturned by the state Supreme Judicial Court after a lengthy court battle. Swift then got the legislature to enlarge the size of the board to five members, and she was able to select a working majority of new members. She chose Matthew J. Amorello, a former state senator and Highway Commissioner, as the Authority's new chairman.

Amorello, like McKinnon before him, understood retail politics and the importance of establishing bipartisan consensus. He quickly took charge of

Turnpike Chairman Matt Amorello with Governor Jane Swift

MassPike and provided the optimism, steady calm, and focus that the agency needed to get the job done. Amorello's temperament, and his style, earned him the confidence of the board's vice chairman, Jordan Levy, and Governor Swift's two new board appointees. He understood his mission was to keep the hold the line on Project costs, to ensure that the public completely understood the Project's operations, and to prepare everyone for the inevitable winding-down process. He was fortunate to have at his side as Project director Michael Lewis, a young engineer who had worked his way up through the system to become a quiet, effective leader—someone who helped maintain the confidence of the business community and the FHWA officials who were concerned that the Project stay on course. Amorello presided over the dedication of the Zakim Bridge, and he opened the north- and southbound tunnel components of the underground artery in 2003.

As Amorello presided over the opening of the main Project tunnels, there was still no political consensus over what they should be named. Governor Weld's decision to name the third harbor tunnel after Boston Red Sox slugger Ted Williams had been met with general enthusiasm, and the legislature's decision to name the Greenway after Rose Kennedy was warmly received. But naming the tunnels, and the great bridge crossing the Charles, proved more

Good riddance: the elevated artery comes down, 2004

The long goodbye: the Central Artery fades into history.

difficult tasks. The naming of the bridge in particular showcased Boston at its worst—a fractious, parochial, small-minded place that often had a difficult time embracing the diversity of a great American city.

Kerasiotes had originally wanted to sell the naming rights for the bridge to Fleet Bank in return for a one million dollar payment to the Project. After his departure, civic leader Peter Meade and developer Arthur Winn joined with political and clerical leaders led by Governor Cellucci and Cardinal Bernard Law to propose that the bridge be named in honor of Leonard P. Zakim, a former head of the Anti-Defamation League and "builder of bridges" among differing ethnic and religious groups in the city. Zakim had suffered through a very public illness before his untimely death, and many in the city saw the opportunity to make a fitting tribute to an important civic leader. Some community groups and some narrow-minded political leaders in neighboring Charlestown objected to the Zakim proposal, and a public debate briefly turned ugly. A compromise was reached: the official name of the structure would be the Leonard P. Zakim Bunker Hill Bridge. As time passed, almost everyone referred to the bridge by the convenient shorthand "Zakim Bridge," and that seemed a fitting end to an unwelcome debate.

There was also a partisan political battle over the naming of the north- and southbound tunnels. Most legislators believed that they should be named in honor of Tip O'Neill, as a small expression of gratitude to the former Speaker for his efforts, but Republican Governor Mitt Romney resisted joining the movement. Shakespeare's question "What's in a name?" was never more apt a query than in Boston in the year 2004. The political stalemate appeared resolved when Congressman Michael Capuano persuaded a Republican colleague to include a provision in the federal transportation reauthorization bill designating the underground tunnels the Tip O'Neill Tunnels.

By 2004, the Project had lasted through five governors, four Turnpike Authority chairmen, and seven state Transportation Secretaries. Amorello was the last in a long line of state transportation leaders entrusted with the responsibility of overseeing the massive Big Dig. Amorello's singular achievement was bringing credibility back to Project leadership. He held the line on construction costs. He chose a former state court judge and a seasoned team of legal professionals to serve as the Project's new cost-recovery team—an effort to ensure state and federal officials that the Project would aggressively move against contractors and designers who had overcharged for their services, or who had failed to perform in a manner consistent with industry standards.

Amorello also took a leadership role in fulfilling the Project's final environmental commitments. The parks systems promised to East Boston, and the Rose Kennedy Greenway, were still in design phase during Amorello's first years in office. He reached out to Boston Mayor Thomas Menino and they jointly developed a design-review process that enhanced public confidence that the final result would be of a high quality. In early 2004, as the city prepared to host the Democratic National Convention, Amorello began the process of removing the elevated highway. There was, in the end, no nostalgia for the old elevated highway. It had never been a welcome addition to the Boston landscape, and one was hard pressed to find a citizen not glad to see it go. The removal process was slow and methodical, keeping noise and disruption to a minimum, a reminder that big things often end not with a bang but a whimper.

CONCLUSION

AT THE DAWN OF THE TWENTIETH CENTURY, James Jackson Storrow might easily have imagined the Charles River as a welcoming place for recreational activities, a calm river basin sparkling in the summer sunlight, filled with sailboats and encircled by hundreds of people seeking a moment's refuge from the bustling city. But even the most vivid of imaginations could not have conjured up the view from the Charles at the beginning of the twenty-first century—a view dominated by the great white cables of the Zakim Bridge, an abstract and giant sailboat itself, a powerful presence framing the inner harbor. The Big Dig rebuilt Boston and reaffirmed its reputation as a "livable" city.

It is a rare occurrence in the history of a city that is has a chance to rebuild itself. Cities like Dresden and Berlin and, to a lesser extent, London rebuilt themselves after experiencing massive destruction during the Second World War. Paris, a "sick, moribund, suffocating" medieval city as late as the 1840s was given a new life in the middle of the nineteenth century by Baron Haussmann and Louis Napoleon. But for the most part great urban centers grow, evolve,

Boston after the Big Dig. The Leonard P. Zakim Bunker Hill Bridge has become Boston's new icon.

and change over time, sometimes losing their past during exuberant bursts of renewal, sometimes integrating past with future.

Boston has had several opportunities to improve itself. The filling of the Back Bay, the creation of the Charles River Basin in the nineteenth century, and the great urban renewal efforts of the "New Boston" movement of the mid-twentieth century, helped to prevent Boston from slipping into a permanent "second class" status among great American cities. Boston always had certain advantages—matchless medical facilities, world-renowned colleges and universities, and a place in American history that was treasured and respected by generations. But in order to be viable as a great urban center—as a livable city that attracts a diverse tapestry of citizens to live and work within its confines—Boston, like most cities, needs something special to distinguish it. The Central Artery/Tunnel Project provided the kind of large-scale investment in the city, both in its transportation infrastructure and in the creation of the acres of new parkland across the city, that comes—if a city is lucky—once in a century.

In the last decades of the twentieth century extraordinary men and women believed in Boston, devoted their lives to rebuilding it, and succeeded in that effort. The fruits of their labors will have a profound and lasting impact on this "city on a hill." Each generation will face its own challenges, each will decide how best to contribute to the community in a way that respects its history and preserves its future. And in the final analysis each generation will be judged by how it improves what it was given, and what it leaves to future generations. By that measure, the men and women of Boston who played roles large and small in the building of the Big Dig will have a prominent and permanent place in its pantheon of heroes.

CHRONOLOGY

1948 Master Highway Plan for the Boston Metropolitan Area is submitted to Governor Robert Bradford.

1950 Central Artery construction begins.

1959 Central Artery is completed.

1968 Massachusetts Turnpike Authority issues report and recommendations on third harbor tunnel cutting through East Boston.

1969 Community activists form Greater Boston Committee on the Transportation Crisis.

1970 Governor Francis Sargent halts construction of Inner Belt.

1974 Michael Dukakis is elected governor of Massachusetts.

1975 Frederick Salvucci is named state Transportation Secretary.

1978 Dukakis loses Democratic primary to Edward J. King; King is elected governor.

1982 Dukakis defeats King in primary "rematch"; Dukakis is elected governor and reappoints Salvucci as Transportation Secretary.

1983 Salvucci begins work on Environmental Impact Statement for Central Artery/Tunnel Project.

1985 Federal Highway Administrator Ray Barnhart writes memorandum expressing initial support for the Project.

1987 Central Artery/Tunnel Project is approved by congressional vote; President Reagan's veto is overturned. Allan McKinnon is named chairman of Massachusetts Turnpike Authority.

1990 Lazard Frères issues report recommending that Turnpike Authority operate and manage the completed Central Artery/Tunnel Project. William Weld is elected Governor. James Kerasiotes is named state Highway Commissioner.

1991 Outgoing Environmental Secretary John DeVillars approves Project's Final Environmental Impact Report. FHWA issues Record of Decision—final federal approval prior to construction. Weld appoints commission to review impacts of proposed Charles River crossing. Third harbor tunnel construction begins.

1995 Legislature designates Turnpike Authority as owner/operator of third harbor tunnel. New tunnel opens December 15.

1996 James Kerasiotes becomes Turnpike Authority chairman in July. MassHighway and MassPike enter into first Project Management Agreement.

1999 Leverett Circle Connector opens to traffic. Project construction activities begin reaching peak.

2000 Kerasiotes discloses over $1 billion in Project cost increases; resigns under pressure.

2002 Governor Swift appoints Matthew Amorello chairman of Turnpike Authority.

2003 Amorello opens Leonard P. Zakim Bunker Hill Bridge and north- and southbound portions of new underground artery.

2004 Amorello leads effort to establish governance structure for Rose Kennedy Greenway. Designs of North End and Chinatown Greenway parcels are completed. Old Central Artery is dismantled.

BIBLIOGRAPHY

I RELIED ON A VARIETY OF SOURCES for this book, including my own memory and knowledge of events, public records (including reports on Project finances issued by the Massachusetts Inspector General and the Federal Highway Administration), official Big Dig publications, reported news accounts, interviews, and my own conversations with a variety of individuals. Congressman Joseph Moakley's quotes come directly from his oral history, a part of the Suffolk University Oral History program. The primary sources for the text were:

Bunting, Bainbridge. *Houses of Boston's Back Bay: An Architectural History, 1840–1917.* Cambridge: The Belknap Press, 1967.

Farrell, John A. *Tip O'Neill and the Democratic Century.* Boston: Little, Brown & Company, 2001.

Haglund, Karl. *Inventing the Charles River.* Cambridge: MIT Press, 2003.

Howells, William Dean. *The Rise of Silas Lapham.* New York: Signet, 1963.

Hughes, Thomas P. *Rescuing Prometheus.* New York: Pantheon Books, 1998.

Kennedy, Lawrence W. *Planning the City upon a Hill: Boston Since 1630.* Amherst: University of Massachusetts Press, 1992.

Luberoff, David, with Alan Altshuler and Christie Baxter. *Mega-Project: A Political History of Boston's Multibillion Dollar Artery/Tunnel Project.* A. Alfred Taubman Center for State and Local Government, Harvard University, 1993.

Lupo, Alan. *Rites of Way: The Politics of Transportation in Boston and the U.S. City.* Boston: Little, Brown & Company, 1971.

Lupo, Alan. "Trouble in Eastie." *Boston Sunday Globe,* September 7, 1969.

Marquand, John P. *The Late George Apley.* New York: Pocket Books, 1971.

McNichol, Dan. *The Big Dig.* New York: Silver Lining Books, 2000.

O'Connor, Thomas H. *Building a New Boston.* Boston: Northeastern University Press, 1993.

O'Connor, Thomas H. *The Hub: Boston Past and Present.* Boston: Northeastern University Press, 2001.

Pearson, Henry Greenleaf. *Son of New England: James Jackson Storrow, 1864–1926.* Boston: Thomas Todd Company, 1932.

Rose, Mark H. *Interstate: Express Highway Politics, 1939–1989.* Rev. ed. Knoxville: University of Tennessee Press, 1990.

Shand-Tucci, Douglass. *Built in Boston: City and Suburb.* Boston: New York Graphic Society, 1978.

ACKNOWLEDGMENTS

ROBERT ALLISON'S SKILLFUL EDITORIAL PEN, and his guidance and support, were essential to the success of this book. So, too, was the careful attention given to my original manuscript by Lynn Walterick—my first, and I hope not my last, copyeditor. Fred Salvucci, Jim Kerasiotes, and Matt Amorello have been friends, mentors, clients, and comrades in the continuing saga of the Big Dig. Were it not for each of these men, I would not have been able to play a continuing role in various aspects of the Big Dig. As part of an ongoing academic project, David Luberoff of the Taubman Center for State and Local Government at Harvard University has written a highly detailed history of the early years of the Big Dig that is an invaluable resource, and was an important resource for this book. A good deal of strong reporting by former *Boston Herald* reporter Laura Brown and *Boston Globe* reporters Peter Howe and Thomas Palmer provides an important contemporaneous account of the Project's ups and downs. Dan MacNichol has put together a variety of popular books that offer fascinating pictorial images and vivid, detailed descriptions of the construction of the Project. Alan Lupo's book *Rites of Way* is an important exploration of the anti-highway movement of the 1960s and 1970s. The records of the Boston College Seminars of the 1950s are a fascinating window on the people who shaped those times, and the planning and economic development beliefs they brought to their work. Finally, I'd like to thank Dennis Rahilly and Cynthia Monahan of the Massachusetts Turnpike Authority/Central Artery Tunnel Project, for their assistance with the wonderful photographs that enliven this book.

INDEX